BIOCHEMISTRY RESEARCH TRENDS

NEW DEVELOPMENTS IN TRYPTOPHAN RESEARCH

BIOCHEMISTRY RESEARCH TRENDS

Additional books in this series can be found on Nova's website
under the Series tab.

Additional e-books in this series can be found on Nova's website
under the e-book tab.

BIOCHEMISTRY RESEARCH TRENDS

NEW DEVELOPMENTS IN TRYPTOPHAN RESEARCH

VICTORIA HAYES
EDITOR

nova publishers
New York

Library of Congress Cataloging-in-Publication Data

ISBN: 978-1-63483-025-6
Library of Congress Control Number: 2015940901

Published by Nova Science Publishers, Inc. † New York

CONTENTS

PREFACE

Tryptophan is not only an essential amino acid but also an effective signaling molecule as an indicator of amino acid sufficiency. Serotonin is biosynthesized from the essential amino acid tryptophan. The first chapter of this book deals with the effects of tryptophan as well as some strategies of treatments that modify serotonin levels on cognitive processes during aging and neurodegenerative diseases. The changes of tryptophan metabolism under immunological challenges and physiological conditions like stress are examined as well. In the third chapter, the authors illustrate the key role of L-tryptophan in metabolic functions with special emphasis on the alterations of circadian rhythms. The remaining chapters of the book focus on new research findings on L-tryptophan synthesis from D-serine; the complex experimental and theoretical studies of the tryptophan molecule in its collisions with low and high energy electrons; and finally, a commentary on a novel tryptophan-selective amino acid transporter that is functionally coupled to ID01-dependent signaling pathways.

Chapter 1 – The neurotransmitter serotonin (5-hydroxytryptamine, 5-HT) plays a role in a variety of physiological processes such as food intake, activity rhythms, sexual behaviour and emotional states as a result of its activity in various structures of the central nervous system. Among this functional heterogeneity, the serotonergic system plays a significant role on cognitive function, in particular by interacting with the cholinergic, glutamatergic, dopaminergic or GABAergic systems. Serotonin release increases in hippocampus during working memory task and extensive evidence suggests that different serotonin receptors have a differential a role in learning and memory. Serotonin is biosynthesized from the essential amino acid tryptophan and the effective amount of serotonin *in vivo* depends on the activity of tryptophan hydroxylase (TPH), the rate-limiting enzyme in the synthesis of serotonin. The TPH activity increases after tryptophan ingestion, which shows that this rate-limiting enzyme is far from being saturated in normal conditions, and shows that the synthesis of serotonin can be modulated by tryptophan ingestion. Increased biodisponibility of tryptophan can increase the amount of serotonin in the brain which can improve cognitive performance. In fact, the greater conversion of tryptophan to serotonin in regions such as the hippocampus is consistent with its role as a key structure involved in cognitive functions such as learning and memory. The synthesis of serotonin decreases with age, partly as a consequence of reduced plasma tryptophan by its degradation to kynurenic acid derivatives; and partly due to the proved susceptibility of TPH enzyme to oxidative damage by reactive oxygen species. In this way, a reduced TPH activity could explain the decrease in serotonin brain in aging. Thus, some studies suggest that treatments aimed at modulating the tryptophan in the brain, as well as modulating the activity of TPH enzyme produce a restorative effect of cognitive functions, which are impaired in aging. Accordingly with the free radical theory of aging, dietary antioxidants manipulations are expected to counteract the rate of production of free radicals and the associated damage. In this manner, the administration of antioxidants may restore TPH activity and thus serotonergic functions in brain. This chapter deals with the effects of tryptophan as well as some strategies or treatments that modify serotonin levels on cognitive processes during aging and neurodegenerative diseases.

Chapter 2 – In a recent perspective, with highlighted role of serotonin in depression, the tryptophan as a precursor of serotonin also became of interest in the pathogenesis of depression. The depletion of tryptophan has been discussed in the etiology of depression. Many studies proposed that serotonin deficiency of depression was caused by the tryptophan shunt towards formation of kynurenine (kynurenine shunt). The inflammatory response system in depression also became of interest. The connection between immune/neuroendocrine system and tryptophan metabolism through kynurenine pathway is considered as important pathophysiology of depression. In this chapter, we reviewed the changes of tryptophan metabolism under immunological challenges and physiological conditions like stress.

Chapter 3 – Mammalian cells cannot synthesize the amino acid tryptophan and dietary intake of tryptophan is crucial in condition of rapid tissue growth and catabolic cellular stress. These situations occur in early life, especially with babies who are born underweight. Tryptophan is metabolized in peripheral tissues such as gastrointestinal tract and liver, as well as by the central nervous system. This amino acid is a precursor of key molecules of circadian cycles and energy homeostasis such as serotonin, melatonin and niacins depending on its time of administration. This review will illustrate the key role of L-tryptophan in metabolic functions with special emphasis on the alterations of circadian rhythms.

Chapter 4 – A historical riddle about chiral homogeneity, the origin of L-dominant amino acids in biological world, has intrigued scientists for many years, though there has been no general consensus as yet to it. Life could not been born without homochirality, that is to say, no homochirality – no life. However, we cannot imagine at all when and how the homochirality was established on early earth. Amino acids (also sugars) are stereochemically too labile to retain their homochiral state in natural environment because racemization pressure is overwhelmingly strong, especially in aqueous solution. An elaborate mechanism is necessary to put biological world into homochirally-stable state. The homochirality of contemporary biological world is fully guaranteed by enzyme enantioselectivity. Since the enzyme enantioselectivity has remained unclear, a deeper understanding of its mechanism is required to unravel all the riddle of the origin of homochirality.

We have so far researched a mechanism of tryptophanase enantioselectivity, with the aim of answering its problem. Tryptophanase degrades L-tryptophan by β-elimination reaction, and also synthesizes L-tryptophan from L-serine and indole by β-replacement reaction by use of its absolute enantioselectivity. On the other hand, it is perfectly inert to D-tryptophan and D-serine at all. However, our previous studies reported that tryptophanase become active on D-tryptophan and D-serine in the presence of diammoniumhydrogen and triammonium phosphate. These salts make the enantioselectivity of tryptophanase flexible. The present research provides triammonium phosphate is the most effective salt among several salts to make the enantioselectivity flexible. A metabolic pathway of L-tryptophan synthesis from D-serine were determined to elucidate how ammonium phosphates produced the flexible enantioselectivity, indicating it was the most significant to form an aldimine bond between amino group of D-serine and pyridoxal 5'-phosphate at the first step of the reaction. Investigation of spectroscopically-based conformational change provided the flexible enantioselectivity was emerged by reversible small conformational change in the presence of triammonium phosphate. We describe these new research findings about L-tryptophan synthesis from D-serine, along with approach to tryptophan research from a new angle.

Chapter 5 – The present chapter is devoted to the complex experimental and theoretical studies of the tryptophan ($C_{11}H_{12}N_2O_2$) molecule fragmentation in its collisions with low (<200 eV) and high (9.5 MeV) energy electrons. The mass-spectrometric technique was used to identify the ionic fragments produced as well as to trace their yield dynamics allowing the appearance energies of the main ionic fragments to be found. The experimental apparatus using a magnetic mass-analyzer has been modified and updated according to the purpose of the present studies. A special attention has been paid to the energy and geometric characteristics of the fragments produced. A number of doubly-charged fragment ions (e.g., $C_8H_5N^{2+}$, $C_9H_7N^{2+}$ and $C_9H_9N^{2+}$) have been studied for the first time. The main emphasis has been given to the problem of evaluating the possible mechanisms of the initial molecule fragmentation using the generalized gradient approximation for the exchange-correlation potential in the density functional theory (DFT). The structures of the parent molecule and its relevant fragments have been optimized globally with no symmetry constraint. The bond orders and bond lengths of the tryptophan molecule have been investigated to find the weaker bonds to be destructed. In order to model the fragmentation processes, the possible fragment anions, cations and fragments with a zero charge both with and without geometry optimization have been evaluated. The dissociation energies have been calculated as the difference between the total energy of the tryptophan molecule and the sum of the energies of the fragments predicted. The geometrical parameters of the initial tryptophan molecule rearrangement have also been analyzed. It has been confirmed that the main channel of the tryptophan molecule dissociation is related to the simple cleavage of the C_α–C_β bond of the initial molecule. This bond dissociation is the most probable when electron energies slightly exceed the tryptophan molecule ionization potential. When the electron energy is sufficient to ionize deeper molecular orbitals, first of all, the lone pair ones of nitrogen and oxygen, dissociation of the O and N atom neighboring bonds takes place, and, as a consequence, more fragmentation channels occur. However, the number of such dissociation channels is not significant because the aromatic group stabilizes the positive charge and reduces the fragmentation. The results obtained are being compared to those found in the relevant databases and literature.

Chapter 6 – Tryptophan is not only an essential amino acid but also an effective signaling molecule as an indicator of amino acid sufficiency. It is catabolized by indoleamine 2,3-dioxygenase 1 (IDO1) with kynurenine as one of the final end products. Induction of IDO1 in cells by a variety of extracellular signals (e.g., interferon-γ, IFN-γ) leads to tryptophan depletion and kynurenine generation inside the cells. IDO1-mediated tryptophan catabolism impacts on three different effector pathways: activation of the protein kinase GCN2 that blocks the protein translation factor eIF-2 thereby blocking protein synthesis, suppression of mTOR thereby interfering with amino acid sufficiency signaling, and activation of the nuclear receptor AhR through kynurenine. As such, IDO1-dependent tryptophan metabolism influences a plethora of signaling pathways. IDO1 induced in one cell type can have an impact on a neighboring cell type through rapid degradation of tryptophan in IDO1-positive cells thereby causing depletion of extracellular tryptophan, thus making this important amino acid unavailable for the neighboring cell type. This is the mechanism responsible for the ability of IDO1-expressing antigen-presenting immune cells to suppress proliferation of effector T cells. An often-overlooked component of this mechanism is the transport process in the plasma membrane of IDO1-positive cells that delivers extracellular tryptophan to the intracellular IDO1. The existence of a transporter selective for tryptophan has been identified in IDO1-expressing cells at the functional level. As expected from the obligatory nature of the transporter for the ability of IDO1 to deplete extracellular tryptophan, the induction of the enzyme and the transporter occurs in a coordinated manner, in most cases, by the same signaling pathways. For example, IFN-γ and AhR that induce IDO1 also induce the tryptophan-selective transporter in the same cell. It is readily apparent that pharmacological blockade of IDO1-mediated tryptophan depletion has therapeutic potential in clinical and pathological conditions that require activation of the immune function as a treatment. Inhibition of either IDO1 or the tryptophan transporter would block this pathway, and both strategies might possibly have comparable therapeutic outcome. Nonetheless, current efforts in this area are almost exclusively focused on IDO1. One of the primary reasons for the apparent lack of interest in the tryptophan transporter as a therapeutic target is that neither the protein nor the gene responsible for the transport process has been identified at the molecular level. But pharmacological targeting of the transporter does not necessarily have to depend on the molecular identity of the transporter. The transport process is known at the functional level; IFNγ-inducible transport function that is selective for tryptophan can be used as a read out for the IDO1-associated tryptophan transporter in the design and development of pharmacological agents to block the transport process. Of course, information on the structure of the transporter protein would facilitate the discovery process in terms of modeling potential inhibitors or blockers of the transporter.

In: New Developments in Tryptophan Research ISBN: 978-1-63483-025-6
Editor: Victoria Hayes

Chapter 1

POSITIVE EFFECTS OF TRYPTOPHAN AND SEROTONIN ON COGNITION DURING AGING AND RELATED-DISEASES

David Moranta, Fiorella Sarubbo, Margarita R. Ramis, Sara Aparicio, Antonio Miralles and Susana Esteban

Laboratory of Neurophysiology, Department of Biology, University of the Balearic Islands, Palma, Spain

ABSTRACT

The neurotransmitter serotonin (5-hydroxytryptamine, 5-HT) plays a role in a variety of physiological processes such as food intake, activity rhythms, sexual behaviour and emotional states as a result of its activity in various structures of the central nervous system. Among this functional heterogeneity, the serotonergic system plays a significant role on cognitive function, in particular by interacting with the cholinergic, glutamatergic, dopaminergic or GABAergic systems. Serotonin release increases in hippocampus during working memory task and extensive evidence suggests that different serotonin receptors have a differential a role in learning and memory.

Serotonin is biosynthesized from the essential amino acid tryptophan and the effective amount of serotonin *in vivo* depends on the activity of tryptophan hydroxylase (TPH), the rate-limiting enzyme in the synthesis of serotonin. The TPH activity increases after tryptophan ingestion, which shows that this rate-limiting enzyme is far from being saturated in normal conditions, and shows that the synthesis of serotonin can be modulated by tryptophan ingestion. Increased biodisponibility of tryptophan can increase the amount of serotonin in the brain which can improve cognitive performance. In fact, the greater conversion of tryptophan to serotonin in regions such as the hippocampus is consistent with its role as a key structure involved in cognitive functions such as learning and memory.

The synthesis of serotonin decreases with age, partly as a consequence of reduced plasma tryptophan by its degradation to kynurenic acid derivatives; and partly due to the proved susceptibility of TPH enzyme to oxidative damage by reactive oxygen species. In this way, a reduced TPH activity could explain the decrease in serotonin brain in aging. Thus, some studies suggest that treatments aimed at modulating the tryptophan in the brain, as well as modulating the activity of TPH enzyme produce a restorative effect of cognitive functions, which are impaired in aging. Accordingly with the free radical theory of aging, dietary antioxidants manipulations are expected to counteract the rate of production of free radicals and the associated damage. In this manner, the administration of antioxidants may restore TPH activity and thus serotonergic functions in brain. This chapter deals with the effects of tryptophan as well as some strategies or treatments that modify serotonin levels on cognitive processes during aging and neurodegenerative diseases.

Physiological Functions of Serotonergic System

The serotonergic system is possibly the most abundant monoaminergic system in the brain, with the highest neuronal population compared to other monoamines. The neurotransmitter serotonin (5-hydroxytryptamine, 5-HT) is a classic neurotransmitter and one of the first described. It is an indolamine which exerts multiple functions through complex regulatory systems. The raphe nuclei, which contain the majority of 5-HT neurones (Figure 1), give rise to projections that are widely distributed throughout the brain (Vertes et al., 1999). Serotonergic neurons of the raphe nuclei project to diverse brain regions that control a wide range of physiological processes such as food intake, sleep, circadian rhythms, sexual behaviour and emotional states, neuroendocrine function, body temperature, pain, motor activity and cognitive functions (Sirvio et al., 1994; Olivier, 2015). Serotonin produces its effects through a variety of cell membrane receptors. In fact, the knowledge of serotonin system has advanced significantly because of the identification, classification and cloning of multiple and diverse 5-HT receptors (Barnes and

Sharp, 1999). With the exception of the ligand-gated ion channel 5-HT3 class, the endogenous agonist, 5-TH, interact with G-protein-coupled receptors (GPCRs) (Leonard, 1992). The diversity of metabotropic 5-HT receptors have been subdivided into families and subtypes that exert their effects through G-protein coupled receptor-activated second messengers as follows: 5-HT$_{1A\text{-}F}$ ($G_{i/o}$), 5-HT$_{2A\text{-}C}$ (G_q), 5-HT$_4$ (G_s), 5-HT$_{5A}$ ($G_{i/o}$), 5-HT$_6$ (G_s), and 5-HT$_7$ (G_s) (Hayes and Greenshaw, 2011; Alexander et al., 2013).

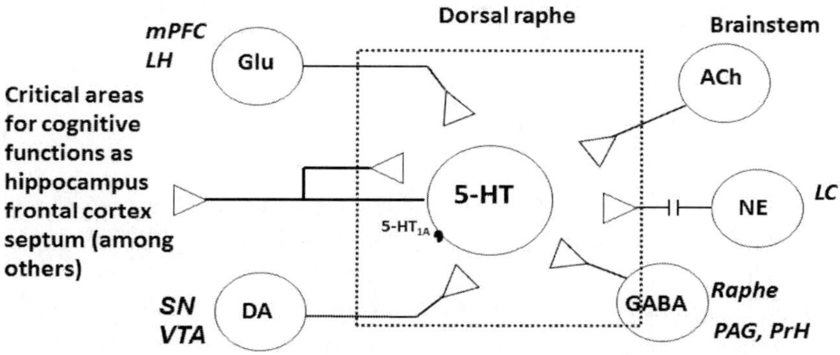

Figure 1. Schematic representation of the main neuronal structures involved in the modulation of serotonin (5-HT) activity in the dorsal raphe nucleus, and location of 5-HT$_{1A}$ inhibitory somatodendritic autoreceptor. Norepinephrine (NE) neurons from the locus coeruleus (LC), GABA local interneurons and GABA afferents from the periaqueductal gray matter (PAG), and the prepositus hypoglossal nucleus (PrH), glutamate (GLU) afferents from the medial prefrontal cortex (mPFC) and lateral habenula (LH), dopamine (DA) neurons from Substantia Nigra (SN) and Ventrotegmental area (VTA), acetylcholine neurons (Ach) from brainstem. Other modulators of raphe activity are neuropeptides as substance P, galanin, CCK, hypocretin-orexin. Serotonergic neurons of the raphe nuclei project to diverse brain regions that control a wide range of physiological processes as cognitive functions; see the main text for specific comments on the in vivo interactions between 5-HT and other neurotransmitter systems.

The serotonergic system has often been described as a modulatory system, characterized by a broad anatomical distribution in the central nervous system (CNS). Because nearly all brain functions were affected by 5-HT (directly or indirectly, e.g., via GABA), it was suggested a lack of functional specialization for this system. Anyway, among its many actions, the serotonergic system plays a significant role in cognitive function, in particular by interacting with the cholinergic, glutamatergic, dopaminergic or GABAergic systems (Cassel and Jeltsch, 1995; Steckler and Sahgal, 1995; Kia et al., 1996).

Time ago, serotonin has been implicated in the processes of learning and memory, partly because the 5-HT receptors have been found in the brain areas involved in learning and memory (Zola-Morgan and Squire, 1993; Meneses and Perez-Garcia, 2007). Nowadays the serotonin is receiving increasing attention and growing evidence suggests that the 5-HT pathways, reuptake transporters, and 5-HT receptors represent a target for the treatment of deficits in learning and memory (Roberts and Hedlund, 2011; Geldenhuys and Van der Schyf, 2011; Rodriguez et al., 2012).

SEROTONIN SYNTHESIS. EFFECTS OF TRYPTOPHAN INTAKE

Serotonin is biosynthesized from the essential amino acid tryptophan and the rate-limiting enzyme in serotonin synthesis is the tryptophan hydroxylase (TPH), which determines the effective amount of serotonin *in vivo*. TPH is known to exist in two isoforms: TPH-1 is mainly expressed in the pineal gland and in gut enterochromaffin cells (Walther et al., 2003), where synthesized 5-HT leads to melatonin synthesis; and TPH-2 is preferentially expressed in the brain where it plays a fundamental role in 5-HT synthesis (Zhang et al,. 2004). TPH-2 activity increases after tryptophan ingestion, which shows that the rate-limiting enzyme is far from being saturated in normal conditions. Thus, serotonin synthesis can be modulated by tryptophan ingestion (Esteban et al., 2004, 2012). At contrary, it is well known that tryptophan depletion reduces the synthesis of the neurotransmitter serotonin in the CNS (Carpenter et al., 1998; Williams et al., 1999; Fadda et al., 2000; Moreno et al., 2010).

Regarding TPH-1 activity, circulating melatonin levels also shows elevation after the tryptophan administration, which is dependent on a dose and time (Hajak et al., 1991; Esteban et al., 2004) and 5-HT-releasing drugs (Huether et al., 1993). In fact, increased tryptophan intake is transformed not only into rises in the quantity of brain serotonin but also in melatonin at pineal and at plasma (Esteban et al., 2004, 2012; Cubero et al., 2006; Garau et al., 2006a; Sanchez et al., 2008). Main differences are observed when tryptophan ingestion arises at the beginning of light or dark phases. During night, tryptophan administration leads to a smaller increase in serotonin, compared with light time administration since it is used as melatonin synthesis substrate. In good agreement with these results, increased tryptophan intake during night produces activity/rest-cycle changes (Garau et al., 2006b). Human studies show a circadian rhythm of tryptophan in breast milk that affects the sleep in newborns (Cubero et al., 2005); which led to design dissociated formulas for infant milk that improve the nocturnal sleep in infants; containing lower levels of tryptophan during the day and higher levels of tryptophan at night (Cubero et al., 2006, 2007; Aparicio et al., 2007).

In general, increased biodisponibility of tryptophan in the brain induces changes in serotonin levels with relevant physiological consequences. Thus, decreasing brain 5HT levels by acute tryptophan depletion has been shown to impair cognition (Haider et al., 2005), whereas the enhanced brain serotonin activity has been shown to improve cognitive performance in animals and humans. In this manner, the increase of serotonin in the brain (and particularly in hippocampus) improved cognitive performance (Haider et al., 2006; Khaliq et al., 2006; Esteban et al., 2010a). In fact, the greater conversion of tryptophan to serotonin in hippocampus is consistent with its key role in learning and memory.

Effects of Aging on Serotonergic System

Several evidences show that serotonergic neurotransmission changes with age. The synthesis of serotonin decreases with age in parallel to an age-dependent decrease in TPH activity (Esteban et al., 2010a,b; Sarubbo et al., 2015). It is partly result of reduced plasma tryptophan by its degradation to kynurenic acid derivatives by enhanced indoleamine (2,3)-dioxygenase (IDO) activity with age (Frick et al., 2004). On the other hand, accordingly with the free radical theory of aging (Harman, 1956), oxidative stress has been indicated in numerous studies such an important contributor for the development and pathogenesis of a wide variety of age-related neurological decline and neurodegenerative diseases (Bishop et al., 2010). In this way, the TPH enzyme has proved to be susceptible to oxidative damage by reactive oxygen species (Kuhn and Arthur, 1996; Hussain and Mitra, 2000). In consequence, a reduced TPH activity could explain the decrease of serotonin in brain aging (Miguez et al., 1999; Tsunemi et al., 2005; Esteban et al., 2010a,b).

While the reduced tryptophan intake impaired learning and memory (van der Stelt et al., 2004; Jenkins et al., 2010), the chronic increase in 5-HT neurotransmission by oral administration of tryptophan was associated with improvements in memory acquisition, consolidation and storage in rodents (Haider et al., 2007). In patients with AD, reduced tryptophan intake further deteriorates cognitive function (Porter et al., 2000). Thus, some studies suggest that treatments aimed at increasing the levels of tryptophan in the brain as well as modulating the activity of TPH enzyme may be accompanied by restorative effect of cognitive functions, which are impaired in aging (Levkovitz et al., 1994; Haider et al., 2006; Khaliq et al., 2006; Esteban et al., 2010a,b; Sarubbo et al., 2015).

Several studies have demonstrated the importance of dietary antioxidants to lower the production rate of free radicals and associated oxidative damage, and thus in promoting health, and in limiting the aging effects (Khurana et al., 2013), in part due to a restoring effect of TPH activity in brain (Esteban et al, 2010a; Moranta et al, 2014; Sarubbo et al., 2015).

In addition, age-related changes on serotonin receptors have also been observed. The 5-HT_{1A} autoreceptor declines in rat cortex, hippocampus and cholinergic basal forebrain nuclei during aging (Nyakas et al., 1997). In humans, post-mortem examinations have detected decreased densities of 5-HT_{1A} receptor proteins expressed in cortex (Tauscher et al., 2001) and especially in hippocampus as age advances (Burnet et al., 1994). It have been hypothesized that these age-related disruptions in serotonergic neurotransmission at discrete brain sites as hippocampus predispose the elderly to memory decline (Luine et al., 1990). Alzheimer's disease (AD) is one of the major neurodegenerative diseases that deteriorates cognitive functions and primarily affects associated brain regions involved in learning and memory, such as the neocortex and the hippocampus. Numerous studies have confirmed the pathophysiological role of the 5-HT system in AD and that several drugs enhancing 5-HT neurotransmission are effective in treating the AD-related cognitive and behavioural deficits (see Rodriguez et al., 2012 for a review).

Effects of Serotonin on Cognitive Processes

Abundant evidence attributed to serotonin a significant role in cognitive processes. Particularly due to the role of several serotonin-receptor subclasses in cognition and memory, and also due to the association of the serotonergic system with brain areas that regulate memory and learning (Meneses and Perez-Garcia, 2007; Geldenhuys and Van der Schyf, 2011). For this reasons serotonin receptors have received increasing attention as potential therapeutic targets for treatment of learning and memory disorders.

Involvement of the serotonergic system in cognitive functions is particularly mediated through its interactions with the cholinergic, dopaminergic, GABAergic or glutamatergic systems (see below). Serotonin action is mediated via specific receptors located in brain structures related to memory/learning functions, as hippocampus and frontal cortex among others (Buhot et al., 2000; Meneses 2003). Regarding this, several studies have shown that certain serotonin receptor ligands have the ability to modify memory/cognition in animal studies. Stimulation of 5-$HT_{2A/2C}$ or 5-HT_4 receptors as well as inhibition of 5-HT_{1A}, 5-HT_3 and 5-HT_{1B} receptors alleviate memory impairment and facilitate learning under conditions that require a high cognitive demand (Buhot et al., 1999, 2000). Moreover, inhibition of 5-$HT_{2A/2C}$ and 5-HT_4 receptors or stimulation of 5-HT_{1A}, 5-HT_3 and 5-HT_{1B} receptors induces an opposite effect (Meltzer et al., 1998; Perez-Garcia and Meneses, 2008). In general, most studies attribute a key role in memory processes at 5-HT_1, 5-HT_4 and 5-HT_6 receptors (King et al., 2008; Geldenhuys and Van der Schyf, 2008, 2009; Glennon et al., 2010). More recently it is also attributed a key role at 5-HT_7 receptors that are localized in brain areas mediating memory but whose role on memory should be explored more extensively (Roberts and Hedlund, 2011; Meneses, 2014).

5-HT_{1A} Receptors

Related to memory process, the 5-HT_{1A} receptors have been the most extensively studied. Autoradiographic studies in rat brain show that 5-HT_{1A} receptors are present in areas typically associated with memory and learning, including the hippocampus, cortex and raphe nuclei (Pazos and Palacios, 1985; Chalmers and Watson, 1991; Pompeiano et al., 1992). The role of 5-HT_{1A} receptors in cognition is complex. Some studies have reported enhanced memory after the administration of the 5-HT_{1A} receptor agonist 8-OH-DPAT (Manuel-Apolinar and Meneses, 2004), but others reported impaired memory functions after its administration (Carli and Samanin, 1992; Lee et al., 1992; Otano et al., 1999; Misane and Ögren, 2000; Egashira et al., 2006). This deficit in memory functions was found to occur by the stimulation of postsynaptic 5-HT_{1A} receptors in the brain (Misane et al., 1998). Note that 5-HT_{1A} receptors are located both pre- and postsynaptically. The 5-HT_{1A} receptors located presynaptically as somatodendritic autoreceptors in raphe nuclei modulate 5-HT release (Figure 1). The stimulation of presynaptic 5-HT_{1A} autoreceptors inhibits the release of serotonin and therefore causes a reduction in the 5-HT transmission in terminal areas as frontal cortex and the locus coeruleus (Verge et al., 1985; Sprouse and Aghajanian, 1986a,b; Sharp et al., 1989). On the other hand, the stimulation of 5-HT_{1A} receptors located postsynaptically modulate the release of other neurotransmitters in non-serotonergic neurons (e.g., cholinergic, glutamatergic, GABAergic). The stimulation of post-synaptic 5-HT_{1A} inhibits cholinergic nerve projections to the hippocampus and frontal cortex as well as modulates the release of dopaminergic, noradrenergic and GABAergic neurotransmitters in brain areas which are strongly implicated in the regulation of learning and memory functions (Barnes and Sharp, 1999; Jacobs and Azmitia, 1992; Meneses, 2003; Luttgen et al., 2005; Madjid et al., 2006; Meneses and Perez-Garcia, 2007; Ogren et al., 2008; Elvander-Tottie et al., 2009). In fact, the administration of selective 5-HT_{1A} receptor antagonists leads to improvement in cognition and memory, presumably due to increased cholinergic neurotransmission (Millan et al., 2004). In experimental studies, depending on the dose of 8-OH-DPAT administrated, it could be activated presynaptic or postsynaptic 5-HT_{1A} receptors producing different effect on memory functions (Haider et al., 2012). Thus, activation of pre-synaptic 5-HT_{1A} receptors by a low dose of its agonist enhanced the memory functions of rats, which could be explained by a less stimulation of GABAergic neurons and hence more release of acetylcholine and/or dopamine. At contrary, memory was improved after a higher dose of the agonist probably acting on post-synaptic 5-HT_{1A} heteroreceptors. This effect may be due to overstimulation of inhibitory GABAergic neurons, which may inhibit the release of acetylcholine and/or dopamine resulting in impairment of memory (Haider et al., 2012). In summary, all these studies indicate a complex role of 5-HT_{1A} receptors in interaction with other neurotransmitter networks in the modulation of learning and memory but suggest that selective drugs for 5-HT_{1A} receptors constitute new therapeutic treatment for memory disorders (Meneses and Perez-Garcia, 2007).

5-HT_4 Receptors

The 5-HT_4 receptor is present in the limbic system (Vilaro et al., 1996) and predominantly in anatomical structures linked to memory and cognition suggesting an active involvement of these receptors in learning and memory (Moser et al., 2002; Manuel-Apolinar et al., 2005). The activation of 5-HT_4 receptors increased the extracellular level of 5-HT (Ge and Barnes, 1996) and improved cognitive function in animal studies in different cognitive tests

(Fontana et al., 1997; Letty et al., 1997; Lamirault and Simon, 2001; Marchetti et al., 2004; Marchetti et al., 2008). These effects could be mediated by the cholinergic system since 5-HT_4 agonists increase acetylcholine release in the rat frontal cortex and hippocampus (Matsumoto et al., 2001; Consolo et al., 1994; Eglen et al., 1995) and also reverse the cognitive impairment induced by scopolamine treatment (Galeotti et al., 1998; Micale et al., 2006).

5-HT_6 Receptors

In addition, the 5-HT_6 receptor has been identified as a target for attenuating cognitive deficits associated with diseases such as Alzheimer's and has been an etiological contributor in cognition deficits in Alzheimer's disease (AD) (Geldenhuys and Van der Schyf, 2008; Meneses, 2014). Thus, antagonists of 5-HT_6 receptor type may have beneficial effects on cognition and memory (Lieben et al., 2005; Geldenhuys and Van der Schyf, 2008, 2009; Marcos et al., 2008).

Serotonergic System in Alzheimer's Disease

Several alterations in the serotonergic system have been also observed in AD. Notable decreases in 5-HT_{1A} and 5-HT_6 receptor densities have been reported in the hippocampus and frontal cortex of AD patients (Lai et al., 2003b; Lorke et al., 2006). The number of 5-HT_4 receptors is also reduced in hippocampus of AD patients (Wong et al., 1996) but not in cortical regions (Lai et al., 2003a).

The 5-HT_4 agonists have been suggested to be a drug target in the treatment of AD as they may offer effective symptomatic treatments for AD. In this way, cholinergic degeneration in the cortex are among the clinical manifestations observed in the patients with AD and 5-HT_4 agonists increase the release of acetylcholine in the rat frontal cortex and hippocampus (see above). Moreover selective activation of 5-HT_4 receptor seems to increase the formation of non-amyloidogenic soluble amyloid precursor protein-alpha, inhibits the generation of β-amyloid peptide, and increases the survival of transgenic neurons in a dose-dependent manner which was counteracted by 5-HT_4 antagonists (Cho and Hu, 2007). In addition, the 5-HT_6 receptor has more recently gained particular interest for the development of anti-AD therapies due to several studies indicating that antagonists at this receptor type may have beneficial effects on cognition and memory (for a review see Geldenhuys and Van der Schyf, 2011). The proposed mechanism of action for serotonin receptors in AD involves activation of serotonin receptors on the cell surface, which initiates a signaling cascade that leads to ERK phosphorylation and activation; ERK appears to increase α-secretase cleavage and may reduce γ-secretase cleavage of amyloid precursor protein (APP) (Cirrito et al., 2011).

Potential Therapies for Restoring the Cognitive Deterioration Associated with Age or Neurodegeneration

Currently they are being devoting much effort in identifying compounds that can restore cognition and memory in AD patients. Present therapy is focused on increasing the loss of acetylcholine, but there remains a continuing loss of cognitive function that needs to be better addressed (Carter et al., 2010).

Searching for new drugs to treat disorders of cognition and memory, the serotonergic system has emerged as a promising target for new therapies (King et al., 2008). Patients with AD show reduced 5-HT neurotransmission, which correlates with the severity of the disease (Mossello et al., 2008). Aggregation of amyloid-β (Aβ) as toxic oligomers and amyloid plaques within the brain appears to be the pathogenic event that initiates Alzheimer's disease (AD) lesions. Treatment with selective serotonin re-uptake inhibitors (SSRI) increases 5-HT neurotransmission, improves cognition in patients with AD, reduces behavioural disturbances associated with AD (Mowla et al., 2007; Mossello et al., 2008) and also reduces amyloid precursor protein translation and pathogenic Aβ peptide secretion hence potentially decreasing Aβ deposition in AD (Payton et al., 2003; Pakaski et al., 2005). Moreover, exposure to 5-HT increased the nonamyloidogenic processing of amyloid precursor protein metabolite in vitro studies (Robert et al., 2001). In mouse model of AD which exhibit severe amyloid pathology, the chronic treatment with SSRI reduced Aβ plaques in the cortex and the hippocampus (Cirrito et al., 2011), reduced both the presence of plaques and neurofibrillary tangles (Nelson et al., 2007) and it has been reported a bi-phasic increase in serotonin transporter–immunoreactive fibres and terminals that develops concomitantly with AD-related neuropathology (Noristani et al., 2010, 2011).

A distinction needs to be drawn between cognitive decline and neurodegeneration. Whereas the decline in cognitive functions can be largely related to disruption of basal forebrain cholinergic pathways (cholinergic hypothesis) (Cummings and Back, 1998), impaired balance between several neurotransmitters has been implicated in the behavioral and psychological symptoms of dementia, with serotonin (5-HT) playing an essential role (Lanctôt et al., 2001; Garcia-Alloza et al., 2004, 2005). Moreover, serotonergic pathways and receptors also are present in brain areas involved in normal and dysfunctional cognitive processes (Flood et al., 1998). For instance, AD apparently is associated with decrements in 5-HT projections from raphe nucleous, in the uptake/transporter complex and in the number of 5-HT_{1A}, 5-HT_{1B}, 5-HT_{1D}, 5-HT_{2A}, 5-HT_{2C} and 5-HT_4 receptors (Meltzer et al., 1998; Meneses, 1998, 1999) and 5-HT_6 receptors (Lorke et al., 2006).

It must be considered that the cognition-enhancing effects of a compound may offer only symptomatic relief, that could not be directly considered as neuroprotective / neurorestorative effect. In this manner, modulators with ability to enhance cognition would likely be of greater benefit if these are administered with the additional therapeutic option represented by neuroprotective compounds. Neuroprotective compounds / neurorestorative are designed to change the course of disease progression, improving function recovery or prevention of cell death. Thus, the increase in 5-HT neurotransmission has been shown to be effective in reducing plaque pathology in AD (see above). In fact, an increased dietary tryptophan intake reduced the accumulation of intracellular β-amyloid in a mouse model of AD probably as a consequence of increased 5-HT neurotransmission (Noristani et al., 2012).

Examples are the 5-HT_6 receptor antagonists who seem to have proneurogenesis properties, and are able to recover lost neurons (Macmillan et al., 2011). Another option is the 5-HT_4 receptor agonists that modify acetylcholine neurotransmission and also seems to play a role in the regulation of non-amyloidogenic soluble amyloid-ß precursor protein (Robert et al., 2001) in addition to modifying acetylcholine neurotransmission (Consolo al., 1994).

Treatments against cognitive decline with age related to serotonergic system may have an additional beneficial effect. In this way, the age-associated cognitive decline is often accompanied and exacerbated by depression (Ogle et al., 2013). It is estimated that approximately 20% of the elderly with cognitive impairment suffer some form of depression (Meltzer et al., 1998). Serotonin is a key element in the pathology of depression pathophysiology of depressive disorder which can be mediated by deregulated hypothalamic-pituitary-adrenal axis and serotonergic activity, as well as compromised neuronal plasticity and neurodegeneration (Foster, 2006; Lupien et al., 2009). Unravelling if depression precedes or follows the age-related cognitive problems is difficult, but finding common treatments for both problems should be addressed (Meltzer et al., 1998; Berton and Nestler, 2006). In this way, TPH-2 has emerged as a novel drug target for the treatment of depressive disorders (Matthes et al., 2010) and prevention of cognitive decline associated with aging (Esteban et al., 2010a,b, 2012; Sarubbo et al., 2015). There are numerous herbal medicines that have been introduced into psychiatric practice due to its high efficiency and lack of side effects (Thachil et al., 2007; Calabrese et al., 2008). One of this exemples is *Polygonum cuspidatum,* a traditional herbal medicine used widely to treat the neuropsychiatric disorders in Asia, with proven efficacy as reported in several pharmacological and clinical studies (Tredici et al., 1999; Chen et al., 2007). Asian cultures have introduced long-time the use of polyphenolic compounds to effectively treat neuropsychiatric and neurodegenerative diseases. Some of them are curcumin, proanthocyanidins, ferulic acid or resveratrol which are products that protect plants from predation and pathogen invasion. Polyphenolic phytochemicals may be particularly attractive for preventing and possibly treating geriatric depression because they modulate HPA axis, stimulate neurotransmission and forms of neural plasticity that include hippocampal neurogenesis (Meltzer et al., 1998; Berton and Nestler, 2006; Foster, 2006; Lupien et al., 2009), and they have also been hypothesized to underlie on age-related cognitive decline (Lazarov et al., 2010). Resveratrol is a non-flavonoid polyphenol that constitutes the main active component of *P. cuspidatum*, but it is also naturally found in several dietary products like grapes, red wine, between others (Bai et al., 2010). It has been demonstrated that resveratrol may alleviate depression because it has pharmacological profiles that inhibit the monoamines catabolism by monoamine oxidase (MAO)-A and MAO-B, but also inhibits the re-uptake of noradrenaline (NA, norepinephrine) and serotonin (5-HT, 5-hydroxytryptamine) into pre-synaptic neurons (Yáñez et al., 2006a,b; Matos et al. 2009a,b; Xu et al., 2010). These effects result in monoamine neurotransmission facilitation and therefore could improve memory processes (Meneses, 1999; Li et al., 2003; Murchison et al., 2004; Kemp and Manahan-Vaughan, 2005, 2008; Lemon and Manahan-Vaughan, 2006; Lemon et al., 2009). Resveratrol improves cognitive function in models of neurodegenerative diseases (Blanchet et al., 2008; Albani et al., 2010; Baur, 2010; Donmez et al., 2010; Wang et al., 2011; Li et al., 2014;

Lofrumento et al., 2014). Recent data point out that long term administration of resveratrol could aid to delay or prevent deficits in monoamines and memory disabilities associated with healthy aging (Sarubbo et al., 2015), effects that were largely due to an increase of 5-HT synthesis through a protective effect of the limiting enzyme TPH against oxidizing agents. Thus, dietary manipulations aimed at reducing the rate of production of free radical reaction damage, could have a potential role for retardation of age-related oxidative events, in accord with the free radical theory of aging.

As tryptophan ingestion can modulate 5-HT synthesis, the 5-HT decrease that normally occurs during ageing could be prevented with diet tryptophan supplementation, and thus some complaints of aging could also be delayed (Esteban et al., 2012).

Moreover, given the low toxicity of ingested tryptophan, an adequate dietary supply of tryptophan would be of paramount importance and might aid to improve some age related degenerative conditions. There are several food sources of tryptophan: eggs, meat (pork, turkey, chicken, beef, lamb, rabbit), fish (cod, salmon, perch, tuna) and shrimp; it is also present in milk and derived products (cheese, yogurt), and cocoa and chocolate among others. This essential amino acid is also detected in several crops (chickpeas, soybeans, sesame seeds, sunflower seeds, mustard seed, oats, wheat flour, rice, quinoa, potatoes, and beans) and fruits such as banana (Chang et al., 1985; Guillén-Casla et al., 2012). In this way, repeated treatment with L-tryptophan, counteracted the age-related descent in serotonin neurotransmission which was accompanied by a restorative effect on the motor and cognitive functions that were impaired as result of aging (Esteban et al., 2010a). In addition, an adequate dietary supply of tryptophan might also aid to improve some age related degenerative conditions (Noristani et al., 2012) since reduced tryptophan intake further deteriorates cognitive function in AD patients (Porter et al., 2000).

ACKNOWLEDGMENTS

This study was supported by grant Programa Pont La Caixa-UIB (7/2014) and by financial contribution of the Govern Balear (ECT 025 09). Sarubbo F. was supported by UIB fellowship, Ramis M.R. by FPU fellowship from Ministerio de Educación, Cultura y Deporte (MECD) from Spain.

REFERENCES

Albani, D; Polito, L; Signorini, A; Forloni, G. Neuroprotective properties of resveratrol in different neurodegenerative disorders. *Biofactors,* 2010, 36, 370-376.

Alexander, SPH; Benson, HE; Faccenda, E; Pawson, AJ; Sharman, JL; Spedding, M; Peters, JA; Harmar, AJ. CGTP Collaborators. The Concise Guide to PHARMACOLOGY 2013/14: G Protein-Coupled Receptors. *Br. J. Pharmacol.*, 2013, 170, 1459–1581.

Aparicio, S; Garau, C; Esteban, S; Nicolau, MC; Rivero, M; Rial, RV. Chrononutrition: Use of dissociated day/night infant milk formulas to improve the development of the wake-sleep rhythms. Effects of tryptophan. *Nutritional Neurosci.*, 2007, 10, 137-143.

Bai, Y; Mao, QQ; Qin, J; Zheng, XY; Wang, YB; Yang, K; Shen, HF; Xie, LP. Resveratrol induces apoptosis and cell cycle arrest of human T24 bladder cancer cells in vitro and inhibits tumor growth in vivo. *Cancer Sci.,* 2010, 101, 488-493.

Barnes, NM; Sharp, T. A review of central 5-HT receptors and their function. *Neuropharmacology,* 1999, 38, 1083-1152.

Baur, JA. Biochemical effects of SIRT1 activators. *Biochim Biophys Acta,* 2010, 1804, 1626–1634.

Berton, O; Nestler, EJ. New approaches to antidepressant drug discovery: beyond monoamines. *Nat. Rev. Neurosci.*, 2006, 7, 137–151.

Bishop, NA; Lu, T; Yankner, BA. Neural mechanisms of ageing and cognitive decline. *Nature,* 2010, 464, 529–535.

Blanchet, J; Longpré, F; Bureau, G; Morissette, M; DiPaolo, T; Bronchti, G; Martinoli, MG. Resveratrol, a red wine polyphenol, protects dopaminergic neurons in MPTP-treated mice. *Prog. Neuropsychopharmacol. Biol. Psychiatry,* 2008, 32, 1243-1250.

Buhot, MC; Malleret, G; Segu, L. Serotonin receptors and cognitive behavior--an update. *IDrugs,* 1999, 2, 426-437.

Buhot, MC; Martin, S; Segu, L. Role of serotonin in memory impairment. *Ann. Med.*, 2000, 32, 210-221.

Burnet, PW; Eastwood, SL; Harrison, PJ. Detection and quantitation of 5-HT1A and 5-HT2A receptor mRNAs in human hippocampus using a reverse transcriptase-polymerase chain reaction (RT-PCR) technique and their correlation with binding site densities and age. *Neurosci. Lett.,* 1994, 178, 85-89.

Calabrese, V; Cornelius, C; Mancuso, C; Pennisi, G; Calafato, S; Bellia, F; Bates, TE; Giuffrida Stella, AM; Schapira, T; Dinkova Kostova, AT; Rizzarelli, E. Cellular stress response: a novel target for chemoprevention and nutritional neuroprotection in aging, neurodegenerative disorders and longevity. *Neurochem. Res.*, 2008, 33, 2444–2471.

Carli, M; Samanin, R. 8-Hydroxy-2-(di-n-propylamino) tetralin impairs spatial learning in a water maze: role of postsynaptic 5-HT1A receptors. *Br. J. Pharmacol.,* 1992, 105, 720–726.

Carpenter, LL; Anderson, GM; Pelton, GH; Gudin, JA; Kirwin, PD; Price, LH; Heninger, GR; McDougle, CJ. Tryptophan depletion during continuous CSF sampling in healthy human subjects. *Neuropsychopharmacology,* 1998, 19, 26–35.

Carter, MD; Simms, GA; Weaver, DF. The development of new therapeutics for Alzheimer's disease. *Clin. Pharmacol. Ther.*, 2010, 88, 475-486.

Cassel, J-C; Jeltsch, H. Serotonergic modulation of cholinergic function in the central nervous system: cognitive implications. *Neuroscience,* 1995, 69, 1-41.

Chalmers, DT; Watson, SJ. Comparative anatomical distribution of 5-HTlA receptor mRNA and 5-HTlA binding in rat brain: a combined in situ hybridization in vitro receptor autoradiographic study. *Brain Res.*, 1991, 561, 51-60.

Chang, SF; Ayres, JW; Sandine, WE. Analysis of cheese for histamine, tyramine, tryptamine, histidine, tyrosine, and tryptophane. *J. Dairy Sci.,* 1985, 68, 2840-2846.

Chen, L-W; Wang, Y-Q; Wei, L-C; Shi, M; Chan, Y-S. Chinese herbs and herbal extracts for neuroprotection of dopaminergic neurons and potential therapeutic treatment of Parkinson's disease. CNS Neurol. *Disord Drug Targets,* 2007, 6, 273-281.

Cho, S; Hu, Y. Activation of 5-HT4 receptors inhibits secretion of beta-amyloid peptides and increases neuronal survival. *Exp. Neurol.*, 2007, 203, 274-278.

Cirrito, JR; Disabato, BM; Restivo, JL; Verges, DK; Goebel, WD; Sathyan, A; Hayreh, D; D'Angelo, G; Benzinger, T; Yoon, H; Kim, J; Morris, JC; Mintun, MA; Sheline, YI. Serotonin signaling is associated with lower amyloid-beta levels and plaques in transgenic mice and humans. *Proc. Natl. Acad. Sci. USA,* 2011, 108, 14968–14973.

Consolo, S; Amaboldi, S; Giorgi, S; Russi, G; Ladinsky, H. 5-HT4 receptor stimulation facilitates acetylcholine release in rat frontal cortex. *Neuroreport*, 1994, 5, 1230-1232.

Cubero, J; Narcis, D; Aparicio, S; Garau, C; Valero, V; Rivero, M; Esteban, S; Rial, R; Rodriguez, AB; Barriga, C. Improved circadian sleep-wake cycle in infants fed a day/night dissociated formula milk. *NeuroEndocrinology Letters,* 2006, 27, 373- 380.

Cubero J, Narciso D, Terrón P, Rial R, Esteban S, Rivero M, Parvez H, Rodríguez AB, Barriga C. Chrononutrition applied to formula milks to consolidate infants sleep/wake cycle. *NeuroEndocrinology Letters* 2007; 28:360-366.

Cubero, J; Valero, V; Sanchez, J; Rivero, M; Parvez, H; Rodríguez, AB; Barriga, C. The circadian rhythm of tryptophan in breast milk affects the rhythms of 6-sulfatoxymelatonin and sleep in newborn. *Neuroendocrinol. Lett.,* 2005, 26, 657–661.

Cummings, JL; Back, C. The cholinergic hypothesis of neuropsychiatric symptoms in Alzheimer's disease. *Am. J. Geriatr Psychiatry*, 1998, 6(Suppl 1), S64-S78.

Donmez, G; Wang, D; Cohen, DE; Guarente, L. SIRT1 suppresses beta-amyloid production by activating the alpha-secretase gene ADAM10. *Cell,* 2010, 142, 320-332.

Egashira, N; Yano, A; Ishigami, N; Mishima, K; Iwasaki, K; Fujioka, M; Matsushita, M; Nishimura, R; Fujiwara, M. Investigation of mechanisms mediating 8-OH-DPAT induced impairment of spatial memory: involvement of 5-HT1A receptors in the dorsal hippocampus in rats. *Brain Res.,* 2006, 1069, 54–62.

Eglen, RM; Wong, EHF; Dumuis, A; Bockaert, J. Central 5-HT4 receptors. *Trends Pharmacol. Sci.*, 1995, 16, 391-398.

Elvander-Tottie, E; Eriksson, TM; Sandin, J, Ogren, SO. 5-HT1A and NMDA receptors interact in the rat medial septum and modulate hippocampal dependent spatial learning. *Hippocampus,* 2009, 19, 1187–1198.

Esteban, S; Garau, C; Aparicio, S; Moranta, D; Barceló, P; Fiol, MA; Rial, RV. Chronic melatonin treatment and its precursor L-tryptophan improve the monoaminergic neurotransmission and related behavior in the aged rat brain. *J. Pineal Res.*, 2010a, 48, 170-177.

Esteban, S; Garau, C; Aparicio, S; Moranta, D; Barceló, P; Ramis, M; Tresguerres, JAF; Rial; RV. Improving effects of long-term growth hormone treatment on monoaminergic neurotransmission and related behavioral tests in aged rats. *Rejuvenation Res.,* 2010b, 13, 707-716.

Esteban, S; Garau, C; Aparicio, S; Nicolau, MC; Rial, RV. Tryptophan Ingestion Improve the Synthesis of Serotonin and Melatonin and May Be Related With Delay of Some Conditions of Aging. In: Tryptophan:

Dietary Sources, Functions and Health Benefits Blake L. Whitley and Sarah H. Thornton (Editors). Series: Human Anatomy. Nova Sciences Publlishers, Inc, 2012, p 39-57.

Esteban, S; Nicolaus, C; Garmundi, A; Rial, RV; Rodríguez, A; Ortega, E; Ibars, CB. Effect of orally administered L-tryptophan on serotonin, melatonin and the innate immune response. *Mol. Cell Biochem.*, 2004, 267, 39-46.

Fadda, F; Cocco, S; Stancampiano, R. A physiological method to selectively decrease brain serotonin release. *Brain Res. Brain Res. Protoc*, 2000, 5, 219–222.

Flood, JF; Farr, SA; Uezu, K; Morley, JE. Aged-related changes in septal serotonergic, GABAergic and glutamatergic facilitation of retention in SAMP8 mice. *Mech. Ageing Dev.*, 1998, 105, 173–188.

Fontana, DJ; Daniels, SE; Wong, EH; Clark, RD; Eglen, RM. The effects of novel, selective 5-hydroxytryptamine (5-HT)4 receptor ligands in rat spatial navigation. *Neuropharmacology,* 1997, 36, 689–696.

Foster, TC. Biological markers of age-related memory deficits: Treatment of senescent physiology. *CNS Drugs,* 2006, 20, 153–166.

Frick, B; Schroecksnadel, K; Neurauter, G; Leblhuber, F; Fuchs, D. Increasing production of homocysteine and neopterin and degradation of tryptophan with older age. *Clin. Biochem.*, 2004, 37, 684–687.

Galeotti, N; Ghelardini, C; Bartolini, A. Role of 5-HT4 receptors in the mouse passive avoidance test. *J. Pharmacol. Exp. Ther.,* 1998, 286, 1115-1121.

Garau, C; Aparicio, S; Rial, RV; Nicolau, MC; Esteban, S. Age-related changes in circadian rhythm of serotonin synthesis in ring doves: Effects of increased tryptophan ingestion. *Experimental Gerontology* 2006a, 41:40-48.

Garau C, Aparicio S, Rial RV, Nicolau MC, Esteban S. Age-related changes in the activity-rest circadian rhythms and c-fos expression of ring Doves with aging. Effects of tryptophan intake. *Exp. Gerontol.*, 2006b, 41, 430-438.

Garcia-Alloza, M; Gil-Bea, FJ; Diez-Ariza, M; Chen, CP; Francis, PT; Lasheras, B; Ramirez, MJ. Cholinergic-serotonergic imbalance contributes to cognitive and behavioral symptoms in Alzheimer's disease. *Neuropsychologia*, 2005, 43, 442-449.

Garcia-Alloza, M; Hirst, WD; Chen, CP; Lasheras, B; Francis, PT; Ramirez, MJ. Differential involvement of 5-HT(1B/1D) and 5-HT6 receptors in cognitive and non-cognitive symptoms in Alzheimer's disease. *Neuropsychopharmacology,* 2004, 29, 410-416.

Ge, J; Barnes, NM. 5-HT4 receptor-mediated modulation of 5-HT release in the rat hippocampus in vivo. *Br. J. Pharmacol.*, 1996, 117, 1475-1480.

Geldenhuys, WJ; Van der Schyf, CJ. Role of serotonin in Alzheimer's disease: a new therapeutic target? *CNS Drugs*, 2011, 25, 765-781.

Geldenhuys, WJ; Van der Schyf, CJ. Serotonin 5-HT6 receptor antagonists for the treatment of Alzheimer's disease. *Curr Top Med Chem*, 2008, 8, 1035-1048.

Geldenhuys, WJ; Van der Schyf, CJ. The serotonin 5-HT6 receptor: a viable drug target for treating cognitive deficits in Alzheimer's disease. *Expert Rev. Neurother,* 2009, 9, 1073-1085.

Glennon, RA; Siripurapu, U; Roth, BL; Kolanos, R; Bondarev, ML; Sikazwe, D; Lee, M; Dukat, M. The medicinal chemistry of 5-HT6 receptor ligands with a focus on arylsulfonyltryptamine analogs. *Curr. Top Med. Chem.* 2010, 10, 579-595.

Guillén-Casla, V; Rosales-Conrado, N; León-González, ME; Pérez-Arribas, LV; Polo-Díez, LM. Determination of serotonin and its precursors in chocolate samples by capillary liquid chromatography with mass spectrometry detection. *J. Chromatogr. A,* 2012, 1232, 158-165.

Haider, S; Khaliq, S; Ahmed, SP; Haleem, DJ. Long term tryptophan administration enhances cognitive performances and increases 5HT metabolism in the hippocampus of female rats. *Amino Acids,* 2006, 31, 421–425.

Haider, S; Khaliq, S; Haleem, DJ. Enhanced serotonergic neurotransmission in the hippocampus following tryptophan administration improves learning acquisition and memory consolidation in rats. *Pharmacol. Rep.*, 2007, 59, 53–57.

Haider, S; Khaliq, S; Tabassum, S; Haleem, DJ. Role of Somatodendritic and Postsynaptic 5-HT1A Receptors on Learning and Memory Functions in Rats. *Neurochem. Res.,* 2012, 37, 2161–2166.

Haider, S; Shameem, S; Ahmed, SP; Perveen, T; Haleem, DJ. Repeated administration of lead decreased brain 5HT metabolism and produces memory deficits in rats. *Cell Mol. Biol. Lett.*, 2005, 10, 669-676.

Hajak, G; Huether, G; Blanke, J; Blömer, M; Freyer, C; Poeggeler, B; Reimer, A; Rodenbeck, A; Schulz-Varszegi, M; Rüther, E. The influence of intravenous L-tryptophan on plasma melatonin and sleep in men. *Pharmacopsychiatry,* 1991, 24, 17-20.

Harman, D. Aging: a theory based on free radical and radiation chemistry. *J. Gerontol.,* 1956, 11, 298-300.

Hayes, DJ; Greenshaw, AJ. 5-HT receptors and reward-related behaviour: A review. *Neurosci. Biobehav. Rev.*, 2011, 35, 1419–1449.

Huether, G; Poeggeler, B; Adler, L; Ruther, E. Effects of indirectly acting 5-HT receptor agonists on circulating melatonin levels in rats. *Eur. J. Pharmacol.*, 1993, 238, 249–255.

Hussain, AM; Mitra, AK. Effect of aging on tryptophan hydroxylase in rat brain: implications on serotonin level. *Drug Metab. Dispos*, 2000, 28, 1038-1042.

Jacobs BL, Azmitia EC. Structure and function of the brain serotonin system. *Physiological Rev* 1992; 72:165–229.

Jenkins, TA; Elliott, JJ; Ardis, TC; Cahir, M; Reynolds, GP; Bell, R; Cooper, SJ. Tryptophan depletion impairs object-recognition memory in the rat: reversal by risperidone. *Behav. Brain Res.*, 2010, 208, 479–483.

Kemp, A; Manahan-Vaughan, D. Beta-adrenoreceptors comprise a critical element in learning-facilitated long-term plasticity. *Cereb. Cortex*, 2008, 18, 1326-1334.

Kemp, A; Manahan-Vaughan, D. The 5-hydroxytryptamine4 receptor exhibits frequency-dependent properties in synaptic plasticity and behavioural metaplasticity in the hippocampal CA1 region in vivo. *Cereb. Cortex,* 2005, 15, 1037-1043.

Khaliq, S; Haider, S; Ahmed, SP; Perveen, T; Haleem, DJ. Relationship of brain tryptophan and serotonin in improving cognitive performance in rats. *Pak J. Pharm. Sci.*, 2006, 19, 11–15.

Khurana, S; Venkataraman, K; Hollingsworth, A; Piche, M; Tai, TC. Polyphenols: Benefits to the Cardiovascular System in Health and in Aging. *Nutrients,* 2013, 5, 3779-3827.

Kia, HK; Brisorgueil, MJ; Daval, G; Langlois, X; Hamon, M; Vergé, D. Serotonin1A receptors are expressed by a subpopulation of cholinergic neurons in the rat medial septum and diagonal band of Broca-a double immunocytochemical study. *Neuroscience,* 1996, 74, 143–154.

King, MV; Marsden, CA; Fone, KC. A role for the 5-HT(lA), 5-HT4 and 5-HT6 receptors in learning and memory. *Trends Pharmacol. Sci.,* 2008, 29, 482-492.

Kuhn, DM; Arthur, RE. Inactivation of brain tryptophan hydroxylase by nitric oxide. *J. Neurochem.*, 1996, 67, 1072-1077.

Lai, MK; Tsang, SW; Francis, PT; Esiri, MM; Hope, T; Lai, OF; Spence, I; Chen, CP. [3H]GR113808 binding to serotonin 5-HT(4) receptors in the postmortem neocortex of Alzheimer disease: a clinicopathological study. *J. Neural. Transm.*, 2003a, 110, 779-88.

Lai, MK; Tsang, SW; Francis, PT; Esiri, MM; Keene, J; Hope, T; Chen, CP. Reduced serotonin 5-HTlA receptor binding in the temporal cortex correlates with aggressive behavior in Alzheimer disease. *Brain Res.*, 2003b, 974, 82-87.

Lamirault, L; Simon, H. Enhancement of place and object recognition memory in young adult and old rats by RS 67333, a partial agonist of 5-HT4 receptors. *Neuropharmacology,* 2001, 41, 844–853.

Lanctôt, KL; Herrmann, N; Mazzotta, P. Role of serotonin in the behavioral and psychological symptoms of dementia. *J. Neuropsychiatry Clin. Neurosci,* 2001, 13, 5-21.

Lazarov, O; Mattson, MP; Peterson, DA; Pimplikar, SW; van Praag, H. When neurogenesis encounters aging and disease. *Trends Neurosci.*, 2010, 33, 569–579.

Lee, EH; Lin, WR; Chen, HY; Shiu, WH; Liang, KC. Fluoxetine and 8-OH-DPAT in the lateral septum enhances and impairs retention of an inhibitory avoidance response in rats. *Physiol. Behav.*, 1992, 51, 681–688.

Lemon, N; Aydin-Abidin, S; Funke, K; Manahan-Vaughan, D. Locus coeruleus activation facilitates memory encoding and induces hippocampal LTD that depends on beta-adrenergic receptor activation. *Cereb. Cortex*, 2009, 19, 2827-2837.

Lemon, N; Manahan-Vaughan, D. Dopamine D1/D5 receptors gate the acquisition of novel information through hippocampal long-term potentiation and long-term depression. *J. Neurosci.*, 2006, 26, 7723-7729.

Leonard, BE. Sub-types of serotonin receptors: biochemical changes and pharmacological consequences. *Int. Clin. Psychopharmaeol.*, 1992, 7, 13-21.

Letty, S; Child, R; Dumuis, A; Pantaloni, A; Bockaert, J; Rondouin, G. 5-HT4 receptors improve social olfactory memory in the rat. *Neuropharmacology,* 1997, 36, 681-687.

Levkovitz, Y; Richter-Levin, G; Segal, M. Effect of 5-hydroxytryptophane on behavior and hippocampal physiology in young and old rats. *Neurobiol Aging*, 1994, 15, 635–641.

Li, J; Feng, L; Xing, Y; Wang, Y; Du, L; Xu, C; Cao, J; Wang, Q; Fan, S; Liu, Q; Fan, F. Radioprotective and antioxidant effect of resveratrol in hippocampus by activating Sirt1. *Int. J. Mol.,* 2014, 15, 5928-5939.

Li, S; Cullen, WK; Anwyl, R; Rowan, MJ. Dopamine dependent facilitation of LTP induction in hippocampal CA1 by exposure to spatial novelty. *Nat. Neurosci.,* 2003, 6, 526-531.

Lieben, CK; Blokland, A; Sik, A; Sung, E; van Nieuwenhuizen, P; Schreiber, R. The selective 5-HT6 receptor antagonist Ro4368554 restores memory performance in cholinergic and serotonergic models of memory deficiency in the rat. *Neuropsychopharmacology*, 2005, 30, 2169-2179.

Lofrumento, DD; Nicolardi, G; Cianciulli, A; De Nuccio, F; La Pesa, V; Carofiglio, V; Dragone, T; Calvello, R; Panaro, MA. Neuroprotective effects of resveratrol in an MPTP mouse model of Parkinson's-like disease: possible role of SOCS-1 in reducing pro-inflammatory responses. *Innate Immun.*, 2014, 20, 249-260.

Lorke, DE; Lu, G; Cho, E; Yew, DT. Serotonin 5-HT2A and 5-HT6 receptors in the prefrontal cortex of Alzheimer and normal aging patients. *BMC Neurosci,* 2006, 7, 36.

Luine, V; Bowling, D; Hearns, M. Spatial memory deficits in aged rats: contributions of monoaminergic systems. *Brain Res.,* 1990, 537, 271-278.

Lupien, SJ; McEwen, BS; Gunnar, MR; Heim, C. Effects of stress throughout the lifespan on the brain, behaviour and cognition. *Nat. Rev. Neurosci.,* 2009, 10, 434-445.

Luttgen, M; Ogren, SO; Meister, B. 5-HT1A receptor mRNA and immunoreactivity in the rat medial septum/diagonal band of Broca-relationships to GABAergic and cholinergic neurons *J. Chem. Neuroanat.,* 2005, 29, 93-111.

Macmillan, KS; Naidoo, J; Liang, J; Melito, L; Williams, NS; Morlock, L; Huntington, PJ; Estill, SJ; Longgood, J; Becker, GL; McKnight, SL; Pieper, AA; De Brabander, JK; Ready, JM. Development of proneurogenic, neuroprotective small molecules. *J. Am. Chem. Soc.,* 2011, 133, 1428-1437.

Madjid, N; Tottie, EE; Luttgen, M; Meister, B; Sandin, J; Kuzmin, A; Stiedl, O; Ogren, SO. 5-Hydroxytryptamine 1A receptor blockade facilitates aversive learning in mice: interactions with cholinergic and gutamatergic mechanism. *J. Pharmacol. Exp. Ther.*, 2006, 316, 581–591.

Manuel-Apolinar, L; Meneses, A. 8-OH-DPAT facilitated memory consolidation and increased hippocampal and cortical cAMP production. *Behav. Brain Res.,* 2004, 148, 179-184.

Manuel-Apolinar, L; Rocha, L; Pascoe, D; Castillo, E; Castillo, C; Meneses, A. Modifications of 5-HT4 receptor expression in rat brain during memory consolidation. *Brain Res.,* 2005, 1042, 73-81.

Marchetti, E; Chaillan, FA; Dumuis, A; Bockaert, J; Soumireu-Mourat, B; Roman, F. Modulation of memory processes and cellular excitability in the dentate gyrus of freely moving rats by a 5-HT4 receptors partial agonist, and an antagonist. *Neuropharmacology,* 2004, 47, 1021-1035.

Marchetti, E; Jacquet, M; Jeltsch, H; Migliorati, M; Nivet, E; Cassel, JC; Roman, FS. Complete recovery of olfactory associative learning by activation of 5-HT4 receptors after dentate granule cell damage in rats. *Neurobiol. Learn Mem.*, 2008, 90, 185-191.

Marcos, B; García-Alloza, M; Gil-Bea, FJ; Chuang, TT; Francis, PT; Chen, CP; Tsang, SW; Lai, MK; Ramirez, MJ. Involvement of an altered 5-HT-{6} receptor function in behavioral symptoms of Alzheimer's disease. *J. Alzheimers Dis.,* 2008, 14, 43-50.

Matos, MJ; Viña, D; Picciau, C; Orallo, F; Santana, L; Uriarte, E. Synthesis and evaluation of 6-methyl-3-phenylcoumarins as potent and selective MAO-B inhibitors. *Bioorg. Med. Chem. Lett.*, 2009a, 19, 5053-5055.

Matos, MJ, Viña, D; Quezada, E; Picciau, C; Delogu, G; Orallo, F; Santana, L; Uriarte, E. A new series of 3-phenylcoumarins as potent and selective MAO-B inhibitors. *Bioorg. Med. Chem. Lett.*, 2009b, 19, 3268-32670.

Matsumoto, M; Togashi, H; Mori, K; Ueno, K; Ohashi, S; Kojima, T; Yoshioka, M. Evidence for involvement of central 5-HT(4) receptors in cholinergic function associated with cognitive processes: behavioral, electrophysiological, and neurochemical studies. *J. Pharmacol. Exp. Ther.,* 2001, 296, 676–682.

Matthes, S; Mosienko, V; Bashammakh, S; Alenina, N; Bader, M. Tryptophan hydroxylase as novel target for the treatment of depressive disorders. *Pharmacology*, 2010, 85, 95-109.

Meltzer, CC; Smith, G; DeKosky, ST; Pollock, BG; Mathis, CA; Moore, RY; Kupfer, DJ; Reynolds, CF 3rd. Serotonin in aging, late-life depression, and Alzheimer's disease: the emerging role of functional imaging. *Neuropsychopharmacology*, 1998, 18, 407-430.

Meneses, A; Perez-Garcia, G. 5-HT1A receptors and memory. *Neuroscience Biobehav Rev*, 2007, 31, 705-727

Meneses, A. 5-HT system and cognition. *Neurosci. Biobehav. Rev.*, 1999, 23, 1111-1125.

Meneses, A. A pharmacological analysis of an associative learning task: 5-HT1 to 5-HT7 receptor subtypes function on a pavlovian/instrumental auto shaped memory. *Learn Mem.*, 2003, 10, 363-372.

Meneses, A. Memory formation and memory alterations: 5-HT6 and 5-HT7 receptors, novel alternative. *Rev. Neurosci.*, 2014, 25, 325-356.

Meneses, A. Physiological, pathophysiological and therapeutic roles of 5-HT systems in learning and memory. *Rev. Neurosci.*, 1998, 9, 1-13.

Micale, V; Leggio, GM; Mazzola, C; Drago, F. Cognitive effects of SL65.0155, a serotonin 5-HT4 receptor partial agonist, in animal models of amnesia. *Brain Res.*, 2006, 1121, 207-215.

Míguez, JM; Aldegunde, M; Paz-Valiñas, L; Recio, J; Sánchez-Barceló, E. Selective changes in the contents of noradrenaline, dopamine and serotonin in rat brain areas during aging. *J. Neural Transm.*, 1999, 106, 1089-1098.

Millan, MJ; Gobert, A; Roux, S; Porsolt, R; Meneses, A; Carli, M; Di Cara, B; Jaffard, R; Rivet, JM; Lestage, P; Mocaer, E; Peglion, JL; Dekeyne, A. The serotonin lA receptor partial agonist S15535 [4-(benzodioxan-5-yl)l-(indan-2-yl)piperazine] enhances cholinergic transmission and cognitive function in rodents: a combined neurochemical and behavioral analysis. *J. Pharmacol. Exp. Ther.*, 2004, 311, 190-203.

Misane, I; Johansson, C; Ögren, SO. Analysis of the 5-HT1A receptor involvement in passive avoidance in the rat. *Br. J. Pharmacol.*, 1998, 125, 499-509.

Misane, I; Ögren, SO. Multiple 5-HT receptors in passive avoidance: comparative studies of p-chloroamphetamine and 8-OH-DPAT. *Neuropsychopharmacology,* 2000, 22, 168–190.

Moranta, D; Barceló, P; Aparicio, S; Garau, C; Sarubbo, F; Ramis, M; Nicolau, C; Esteban, S. Intake of melatonin increases tryptophan hydroxylase type 1 activity in aged rats: Preliminary study. *Exp. Gerontol.,* 2014, 49, 1-4.

Moreno, FA; Parkinson, D; Palmer, C; Castro, WL; Misiaszek, J; El Khoury, A; Mathé, AA; Wright, R; Delgado, PL. CSF neurochemicals during tryptophan depletion in individuals with remitted depression and healthy controls. *Eur. Neuropsychopharmacol.*, 2010, 20, 18-24.

Moser, PC; Bergis, OE; Jegham, S; Lochead, A; Duconseille, E; Terranova, JP; Caille, D; Berque-Bestel, I; Lezoualc'h, F; Fischmeister, R; Dumuis, A; Bockaert, J; George, P; Soubrié, P; Scatton, B. SL65.0155, a novel 5-hydroxytryptamine(4) receptor partial agonist with potent cognition-enhancing properties. *J. Pharmacol. Exp. Ther.*, 2002, 302, 731-741.

Mossello, E; Boncinelli, M; Caleri, V; Cavallini, MC; Palermo, E; Di Bari, M; Tilli, S; Sarcone, E; Simoni, D; Biagini, CA; Masotti, G; Marchionni, N. Is antidepressant treatment associated with reduced cognitive decline in Alzheimer's disease? *Dement Geriatr Cogn Disord*, 2008, 25, 372-379.

Mowla, A; Mosavinasab, M; Haghshenas, H; Borhani Haghighi, A. Does serotonin augmentation have any effect on cognition and activities of daily living in Alzheimer's dementia? A double-blind, placebo-controlled clinical trial *J. Clin. Psychopharmacol.*, 2007, 27, 484-487.

Murchison, CF; Zhang, XY; Zhang, WP; Ouyang, M; Lee, A; Thomas, SA. A distinct role for norepinephrine in memory retrieval. *Cell,* 2004, 117, 131-143.

Nelson, RL; Guo, Z; Halagappa, VM; Pearson, M; Gray, AJ; Matsuoka, Y; Brown, M; Martin, B; Iyun, T; Maudsley, S; Clark, RF; Mattson, MP. Prophylactic treatment with paroxetine ameliorates behavioral deficits and retards the development of amyloid and tau pathologies in 3xTgAD mice. *Exp. Neurol.*, 2007, 205, 166-176.

Noristani, HN; Meadows, RS; Olabarria, M; Verkhratsky, A; Rodriguez, JJ. Increased hippocampal CA1 density of serotonergic terminals in a triple transgenic mouse model of Alzheimer's disease: an ultrastructural study. *Cell Death Dis.,* 2011, 2, e210.

Noristani, HN; Olabarria, M; Verkhratsky, A; Rodríguez, JJ. Serotonin fibre sprouting and increase in serotonin transporter immunoreactivity in the CA1 area of hippocampus in a triple transgenic mouse model of Alzheimer's disease. *Eur. J. Neurosci.*, 2010, 32, 71-79.

Noristani, HN; Verkhratsky, A; Rodríguez, JJ. High tryptophan diet reduces CA1 intraneuronal β-amyloid in the triple transgenic mouse model of Alzheimer's disease. *Aging Cell*, 2012, 11, 810-822.

Nyakas, C; Oosterink, BJ; Keijser, J; Felszeghy, J; de Jong, GI; Korf, J; Luiten, PGM. Selective decline of 5-HT1A receptor binding sites in rat cortex, hippocampus and cholinergic basal forebrain nuclei during aging. *J. Chem. Neuroanat*, 1997, 13, 53-61.

Ogle, WO; Speisman, RB; Ormerod, BK. Potential of Treating Age-Related Depression and Cognitive Decline with Nutraceutical Approaches: A Mini-Review. *Gerontology,* 2013, 59:23–31.

Ogren, SO; Eriksson, TM; Elvander-Tottie, E; D'Addario, C; Ekstrom, JC; Svenningsson, P; Meister, B; Kehr, J; Stiedl, O. The role of 5-HT1A receptors in learning and memory. *Behav. Brain Res.*, 2008, 195, 54-77.

Olivier, B. Serotonin: A never-ending story. *Eur. J. Pharmacol.*, 2015, 753, 2-18.

Otano, A; Garcia-Osta, A; Ballaz, S; Frechilla, D; Del Rio, J. Facilitation by 8-OH-DPAT of passive avoidance performance in rats after inactivation of 5-HT1A receptors. *Br. J. Pharmacol.*, 1999, 128, 1691-1698.

Pakaski, M; Bjelik, A; Hugyecz, M; Kasa, P; Janka, Z; Kalman, J. Imipramine and citalopram facilitate amyloid precursor protein secretion in vitro. *Neurochem. Int.,* 2005, 47, 190-195.

Payton, S; Cahill, CM; Randall, JD; Gullans, SR; Rogers, JT. Drug discovery targeted to the Alzheimer's APP mRNA 5'-untranslated region: the action of paroxetine and dimercaptopropanol. *J. Mol. Neurosci.*, 2003, 20, 267-275.

Pazos, A; Palacios, JM. Quantitative autoradiographic mapping of serotonin receptors in the rat brain. I. Serotonin 1 receptors. *Brain Res.*, 1985, 346, 205-230.

Perez-Garcia, G; Meneses, A. Memory formation, amnesia, improved memory and reversed amnesia: 5-HT role. *Behav. Brain Res.,* 2008, 195, 17-29.

Pompeiano, M; Palacios, JM; Mengod, G. Distribution and cellular localization of mRNA coding for 5-HT1A receptor in the rat brain: correlation with receptor binding. *J. Neurosci.*, 1992, 12, 440-453.

Porter, RJ; Lunn, BS; Walker, LL; Gray, JM; Ballard, CG; O'Brien, JT. Cognitive deficit induced by acute tryptophan depletion in patients with Alzheimer's disease. *Am. J. Psychiatry*, 2000, 157, 638-640.

Robert, SJ; Zugaza, JL; Fischmeister, R; Gardier, AM; Lezoualc'h, F. The human serotonin 5-HT4 receptor regulates secretion of non-amyloidogenic precursor protein. *J. Biol. Chem.*, 2001, 276, 44881-44888.

Roberts, AJ; Hedlund, PB. The 5-HT(7) receptor in learning and memory. *Hippocampus,* 2011, 22, 762-771.

Rodríguez, JJ; Noristani, HN; Verkhratsky, A. The serotonergic system in ageing and Alzheimer's disease. *Prog. Neurobiol.*, 2012, 99, 15-41.

Sanchez, S; Paredes, SD; Sanchez, CL; Barriga, C; Reiter, RJ; Rodriguez, AB. Tryptophan administration in rats enhances phagocytic function and reduces oxidative metabolism. *Neuro. Endocrinol. Lett.*, 2008, 29, 1026-1032.

Sarubbo, F; Ramis, MR; Aparicio, S; Ruiz, L; Esteban, S; Miralles, A; Moranta, D. Improving effect of chronic resveratrol treatment on central monoamines synthesis and cognition in aged rats. *Age*, 2015 (in press).

Sharp, T; Bramwell, SR; Hjorth, S; Grahame-Smith, DG. Pharmacological characterization of 8-OH-DPAT-induced inhibition of rat hippocampal 5-HT release in vivo as measured by microdialysis. *Br. J. Pharmacol.*, 1989, 98, 989-997.

Sirvio, J; Riekkinen Jr, P; Jakala, P; Riekkinen, PJ. Experimental studies on the role of serotonin in cognition. *Prog Neurobiol*, 1994, 43, 363-379.

Sprouse, JS; Aghajanian, GK. (-)-Propranolol blocks the inhibition of serotonergic dorsal raphe cell firing by 5-HT1A selective agonists. *Eur. J. Pharmacol.,* 1986a, 128, 295-298.

Sprouse, JS; Aghajanian, GK. Responses of hippocampal pyramidal cells to putative serotonin 5-HT1A and 5-HT1B agonists: a comparative study with dorsal raphe neurons. *Neuropharmacology,* 1986b, 27, 707-715.

Steckler, T; Sahgal, A. The role of serotonergic-cholinergic interactions in the_mediation of cognitive behaviour. *Behav. Brain Res.,* 1995, 67, 165-199.

Tauscher, J; Verhoeff, NP; Christensen, BK; Hussey, D; Meyer, JH; Kecojevic, A; Javanmard, M; Kasper, S; Kapur, S. Serotonin 5-HT1A receptor binding potential declines with age as measured by [11C]WAY-100635 and PET. *Neuropsychopharmacology,* 2001, 24, 522-530.

Thachil, AF; Mohan, R; Bhugra, D. The evidence base of complementary and alternative therapies in depression. *J. Affect Disord,* 2007, 97, 23-35.

Tredici, G; Miloso, M; Nicolini, G; Galbiati, S; Cavaletti, G; Bertelli, A. Resveratrol, map kinases and neuronal cells: might wine be a neuroprotectant?. *Drugs Exp. Clin. Res.*, 1999, 25, 99-103.

Tsunemi, A; Utsuyama, M; Seidler, BK; Kobayashi, S; Hirokawa, K. Age-related decline of brain monoamines in mice is reversed to young level by Japanese herbal medicine. *Neurochem. Res.*, 2005, 30, 75-81.

van der Stelt, HM; Broersen, LM; Olivier, B; Westenberg, HG. Effects of dietary tryptophan variations on extracellular serotonin in the dorsal hippocampus of rats. *Psychopharmacology,* 2004, 172, 137-144.

Verge, D; Daval, G; Patey, A; Gozlan, H; el Mestikawy, S; Hamon, M. Presynaptic 5-HT autoreceptors on serotonergic cell bodies and/or dendrites but not terminals are of the 5-HT1A subtype. *Eur. J. Pharmacol.,* 1985, 113, 463-464.

Vertes, RP; Fortin, WJ; Crane, AM. Projections of the median raphe nucleus in the rat. *J. Comp. Neurol.*, 1999, 407, 555-582.

Vilaro, MT; Cortes, R; Gerald, C; Branchek, TA; Palacios, JM; Mengod, G. Localization of 5-HT4 receptor mRNA in rat brain by in situ hybridization histochemistry. *Brain Res. Mol. Brain Res., 1996*, 43, 356-360.

Walther, DJ; Peter, JU; Bashammakh, S; Hörtnagl, H; Voits, M; Fink, H; Bader, M. Synthesis of serotonin by a second tryptophan hydroxylase isoform. *Science,* 2003, 299, 76.

Wang, Y; Xu, H; Fu, Q; Ma, R; Xiang, J. Protective effect of resveratrol derived from Polygonum cuspidatum and its liposomal form on nigral cells in parkinsonian rats. *J. Neurol. Sci.*, 2011, 304, 29-34.

Williams, WA; Shoaf, SE; Hommer, D; Rawlings, R; Linnoila, MI. Effects of acute tryptophan depletion on plasma and cerebrospinal fluid tryptophan and 5-hydroxyindoleacetic acid in normal volunteers. *J. Neurochem.*, 1999, 72, 1641-1647.

Wong, EHF; Reynolds, GP; Bonhaus, DW; Hsu, S; Eglen, RM. Characterization of [SHIGR 113808 binding to 5-HTl receptors in brain tissues from patients with neurodegenerative disorders. *Behav. Brain Res.,* 1996, 73, 249-252.

Xu, Y; Wang, Z; You, W; Zhang, X; Li, S; Barish, PA; Vernon, MM; Du, X; Li, G; Pan, J; Ogle, WO. Antidepressant-like effect of trans-resveratrol: involvement of serotonin and noradrenaline System. *Eur. Neuropsychopharmacol.,* 2010, 20, 405-413.

Yáñez, M; Fraiz, N; Cano, E; Orallo, F. (−)-Trans-epsilonviniferin, a polyphenol present in wines, is an inhibitor of noradrenaline and 5-hydroxytryptamine uptake and of monoamine oxidase activity. *Eur. J. Pharmacol.,* 2006b, 542, 54-60.

Yáñez, M; Fraiz, N; Cano, E; Orallo, F. Inhibitory effects of cis- and trans-resveratrol on noradrenaline and 5-hydroxytryptamine uptake and on monoamine oxidase activity. *Biochem. Biophys. Res. Commun.*, 2006a, 344, 688-695.

Zhang, X; Beaulieu, JM; Sotnikova, TD; Gainetdinov, RR; Caron, MG. Tryptophan hydroxylase-2 controls brain serotonin synthesis. *Science,* 2004, 305, 217.

Zola-Morgan, S; Squire, LR. Neuroanatomy of memory. *Ann. Rev. Neurosci.,* 1993, 168, 547-563.

In: New Developments in Tryptophan Research ISBN: 978-1-63483-025-6
Editor: Victoria Hayes

Chapter 2

KYNURENINE AND SEROTONIN PATHWAYS OF TRYPTOPHAN METABOLISM: THE ETIOLOGY AND PATHOGENESIS OF DEPRESSION

Sang Won Jeon and Yong-Ku Kim*
Department of Psychiatry, College of Medicine, Korea University, Ansan Hospital, Republic of Korea

In a recent perspective, with highlighted role of serotonin in depression, the tryptophan as a precursor of serotonin also became of interest in the pathogenesis of depression. The depletion of tryptophan has been discussed in the etiology of depression. Many studies proposed that serotonin deficiency of depression was caused by the tryptophan shunt towards formation of kynurenine (kynurenine shunt). The inflammatory response system in depression also became of interest. The connection between immune/ neuroendocrine system and tryptophan metabolism through kynurenine pathway is considered as important pathophysiology of depression. In this chapter, we reviewed the changes of tryptophan metabolism under immunological challenges and physiological conditions like stress.

* Corresponding author: Yong-Ku Kim, M.D., Ph.D. Department of Psychiatry, College of Medicine, Korea University, Ansan Hospital, 123 Jeokgeum-ro, Danwon-gu, Ansan-si, Gyeonggi-do, 425-707, Republic of Korea, Tel: +82-31-412-5140, Fax: +82-31-412-5144, E-mail: yongku@korea.ac.kr, yongku0304@gmail.com.

1. Kynurenine and Serotonin Pathways of Tryptophan Metabolism

The balance and interaction of endocrine in the brain and immune system is considered an important factor of pathogenesis in major depressive disorder. The metabolism of tryptophan (TRP) plays an important role in such process. TRP is an essential amino acid, and its metabolism consists of two pathways: 1) the kynurenine (KYN) pathway [1] and 2) the serotonin (5-HT) pathway [2] (Figure 1). TRP, which is mostly consumed via the diet, is known to be metabolized through two pathways [3]. Among them, the KYN pathway is a key pathway in explaining the pathogenesis of depression. The KYN pathway is initiated by the enzyme in indoleamine 2, 3-dioxygenase (IDO).

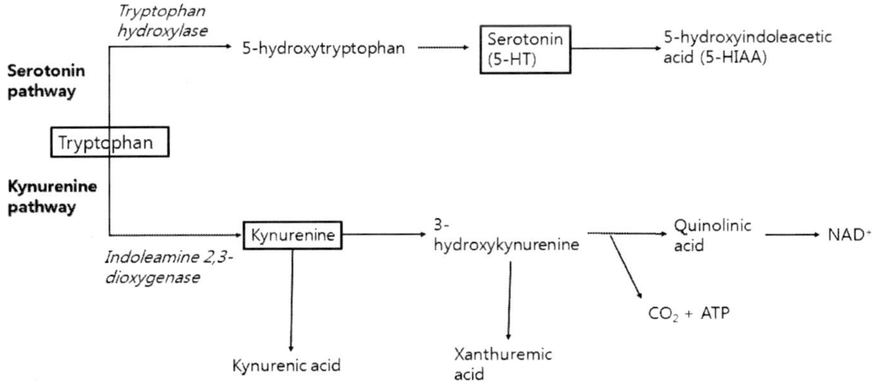

5-HT: serotonins, 5-HIAA: 5-hydroxyindoleacetic acid, ATP: Adenosine triphosphate, NAD: nicotinamide adenine dinucleotide.

Figure 1. Kynurenine and serotonin pathways of tryptophan metabolism.

This IDO is considered to have the greatest effect on the progression of depression. This is because the pro-inflammatory cytokines (IL-1, IFN-α, IFN-γ, and TNF-α) that are considered closely related to depression activate IDO (Figure 2). Activated IDO metabolizes TRP to KYN. When this KYN pathway is activated, it competes with the 5-HT pathway that metabolizes TRP to 5-hydroxytryptophan through tryptophan hydroxylase (TPH). As such, the activated KYN pathway depletes TRP and debases the 5-HT pathway [4]. The depletion of TRP rapidly lifts down the mood and causes vulnerability to depression [5]. In the 1980s, a study was conducted to investigate the correlation between the amount of KYN metabolite and depression. KYN is converted to xanthurenic acid (XA) by 3-hydroxykyurenine (3-OH KYN) (Figure 1). The relationship between the urine secretion of XA and depression was analyzed. The depressed subjects who consumed TRP showed greater XA excretion than the healthy subjects, and their XA excretion improved when their depression improved [6]. In summary, the secretion of pro-inflammatory cytokine activates IDO, the activated IDO interrupts 5-HT synthesis, and 5-HT deficiency causes depression. This theory is well-known as the monoamine theory, which is considered the cause of depression.

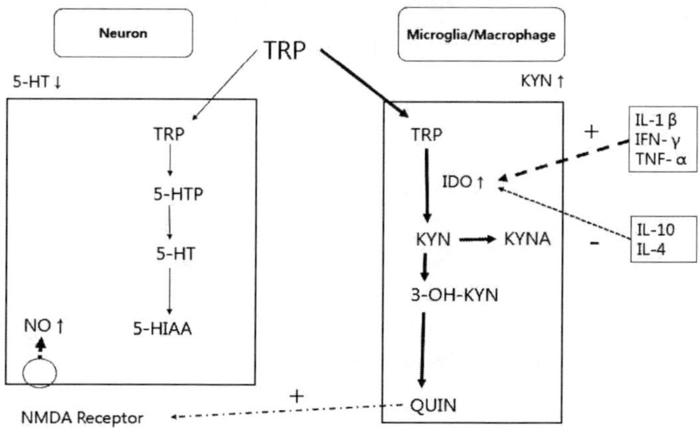

+: activation, -: inhibition, 3-OH KYN: 3-hydroxykynurenine, 5-HT: serotonin, 5-HIAA: 5-hydroxyindoleacetic acid, 5-HTTP: 5-hydroxy tryptophan, IDO:indoleamine 2,3-dioxygenase, IFN: interferon, IL: interleukin, KYN: kynurenine, KYNA: kynurenic acid, NMDA: N-methyl-D-aspartate, NO: nitric oxide, QUIN: quinolinic acid, TNF: tumor necrosis factor, TRP: tryptophan

Figure 2. Tryptophan metabolism and immunological activation in the brain.

2. Immunological Activation and Tryptophan Metabolism in Depression

Another important pathogenesis of depression, along with monoamine hypothesis, is hypothalamic-pituitary-adrenal (HPA) axis hyperactivity hypothesis [7]. This theory considers the role of immune system important and is related to tryptophan metabolism. Monocyte-T-lymphocyte hypothesis was suggested to explain the correlation between major depression and immunological activity [8, 9]. According to the theory, activated macrophages elevate the IL-1 level, and in turn, the elevated IL-1 level stimulates the corticotrophin-releasing factor (CRF) in the paraventricular nucleus (PVN). Also, IL-6 directly works on the hypothalamus and disturbs the secretion of CRF, which results in HPA-axis abnormality [9]. As such, the HPA-axis hyperactivity hypothesis explains the relationship of the HPA-axis hyperactivity and depression and comprehensively explains the role of immunological activation. These two hypotheses (monoamine hypothesis and HPA-axis hyperactivity hypothesis) are core hypotheses that explain the etiology and pathogenesis of major depression.

In the KYN pathway, 3-OH KYN and quinolinic acid (QUIN) show neurotoxic activity, and kynurenic acid (KYNA) shows neuroprotective activity [10]. The imbalance of KYN metabolites with neurotoxic-protective activation are known to damage the hippocampus. A damaged hippocampus is often observed in the patients with chronic depression. In mice studies, activated IDO from systemic immune activation increased the concentration of QUIN in the CSF. On the contrary, the KYNA concentration in the CSF remained the same regardless of the immune activation. Thus, the immune activation increased the QUIN/KYNA ratio in the CSF [11]. In other words, the neurotoxicity of the brain increased. In human studies on IFN-α treatment, the increased serum KYN/KYNA ratio was associated with the depression severity [12]. The depressed subjects showed decreased plasma KYNA and an increased KYN/KYNA ratio.

3. Brain Tryptophan Level and Stress

Repetitive and persistent exposure to stress is considered a cause of depression. As such, the observation of TRP metabolism changes could be one of the methods of approaching the neurobiology of depression.

Immunological challenges or stress is known to elevate the brain TRP level. In animal studies, the brain tryptophan level rose when the animals were subjected to stress such as to immobilization stress, food deprivation, or electric shock [13]. The same level also rose through HPA axis activation when immunological challenges (e.g., IL-1 or lipopolysaccharide(LPS)) were used [14]. The sympathetic nerve system is known to elevate the brain TRP level [15]. However, a study of adrenalectomized mice did not show any change in the brain TRP when the animals were subjected to stress, which confirmed the non-dependence of TRP on adrenocortical activation [15].

Two mechanisms are involved in the determination of the brain TRP level (Figure 3). In the first mechanism, TRP, being an essential amino acid, mostly binds to albumin in plasma. There are still some free TRPs. On the other hand, free fatty acid (FFA), which is sensitively affected by the sympathetic nerve system, is known to be closely associated with the brain TRP level. Under stress, the sympathetic system activates that FFA binds to albumin [16]. This decreases the affinity of TRP to albumin. In other words, the activated sympathetic system causes FFA to bind more to albumin while interrupting the binding of TRP to albumin (Figure 3). As a result, the free TRP level rises in plasma [17]. The second mechanism is related to large neutral amino acid (LNAA). LNAA is located in uptake sites of blood brain barrier (BBB) and consists of aromatic amino acid and branched chain amino acid. TRP and LNAA also compete with each other. The TRP/LNAA ratio in plasma is known to be related to the TRP uptake activity in BBB. An elevated TRP/LNAA ratio increases the brain uptake of TRP and the brain TRP level [16, 17].

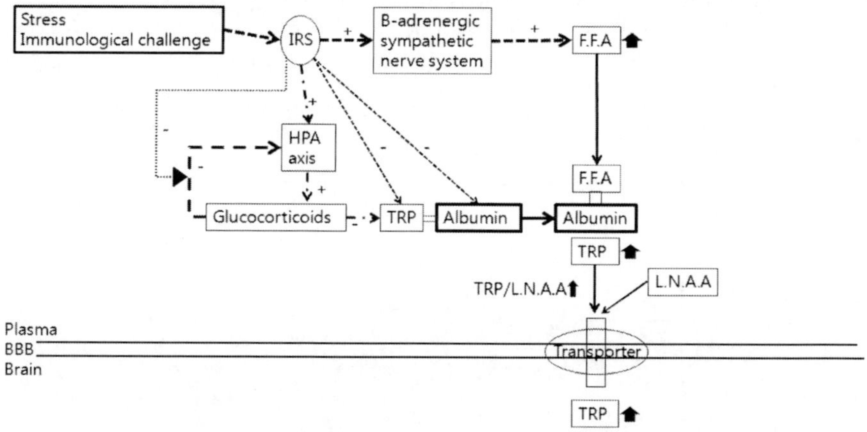

+: activation, -: inhibition, BBB: blood–brain barrier, FFA: free fatty acid, HPA axis: hypothalamo-pituitary-adrenal axis, IRS: inflammatory response system, LNAA: large neutral amino acid, TRP: tryptophan

Figure 3. Uptake of tryptophan in the brain.

In summary, stress elevates the brain TRP level. Even if such was confirmed in mice studies, the TRP uptake is assumed to be elevated under psychological stress in humans. The change in the brain function observed in chronic depression is associated with chronic stress. As such, the mechanism of the change in the brain TRP level under stress can explain the pathology of depression.

4. Tryptophan Metabolism and Stress

Stress is estimated to change TRP metabolism, assuming that it activates IDO and the KYN pathway (Figure 4) [18]. This mechanism was also observed in microglia and astrocytes, as well as in brain neurons [19]. Under stress, mice showed elevated brain TRP, KYN, KYNA, and 3-OH KYN levels. The brain TRP level was 10-40 nmol/g, and the brain KYN level, 400-1200 pmol/g. The KYNA level was 20-40 pmol/g, and the 3-OH KYN level, 40-80 pmol/g [20]. As such, the brain TRP level was greater than the brain KYN, KYNA, and 3-OH KYN levels. Stress increases the TRP uptake in BBB and elevates the brain TRP level. Increased brain TRP compensates for TRP deficiency which is caused by TRP's converting to KYN by IDO activation.

An animal study confirmed a consistent KYN/TRP ratio under stress due to increased both KYN and TRP levels [21]. As such, brain IDO activity is very important under stress, regardless of the immunological challenge. Stress is known to decrease the 5-HT/TRP ratio. The 5-HT pathway is assumed to shift to the KYN pathway under stress. However, not enough studies have been conducted on stress and brain TRP metabolism, so more future studies are required.

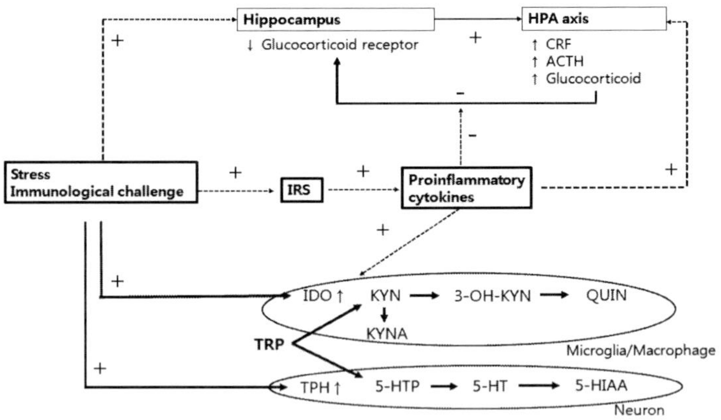

+: activation, -: inhibition, 3-OH KYN: 3-hydroxykynurenine, 5-HIAA: 5-hydroxyindoleacetic acid, 5-HT: serotonin, 5-HTTP: 5-hydroxy tryptophan, ACTH: adrenocorticotropic hormone, CRF: corticotropin releasing factor, HPA axis: hypothalamo-pituitary-adrenal axis, IDO: indoleamine 2,3-dioxygenase, IRS: inflammatory response system, KYN: kynurenine, KYNA: kynurenic acid, QUIN: quinolinic acid, TRP: tryptophan, TPH: tryptophan hydroxylase.

Figure 4. KYN pathway and HPA axis by immunological challenges and stress.

5. Pro-Inflammatory Cytokines and Stress

Stress increases the production of pro-inflammatory cytokines in humans [22]. The human brain and the immune system make up the communication network and interact with each other. The peripheral immune system also interacts with the brain [23]. The pro-inflammatory cytokines secreted from macrophages send signals to the brain through humoral and neural routes. Pro-inflammatory cytokines are large molecules that are hydrophilic, which keeps them from passing through BBB.

Also, they can send signals to the brain using circumventricular organs (CVOs) or the choroid plexus. IL-1b acts as a secondary messenger in CVOs and produces prostaglandin E2 (PGE2). PGE2 is a small molecule that is lipophilic. As such, it can freely diffuse to the brain [24]. Pro-inflammatory cytokines can also use the active transport mechanism to send signals to the brain. Such mechanism has been proven for IL-1a, IL-1b, IL-6, and TNF-α [25]. IL-1b works on glial cells in CVOs and the choroid plexus, and stimulates its own production. As such, IL-1b produces new IL-1b in glial cells using the paracrine way.

In this way, IL-1b can send signals to distant regions [24]. Pro-inflammatory cytokines also send signals to the brain using a neural route, mainly the vagus nerve. The IL-1b secreted in the afferent terminal of the vagus nerve increases the nerve impulse [24]. When the immune system is stimulated, IL-1b also activates the HPA axis and the sympathetic nerve system, thus affecting the brain [24].

Neurochemical change under stress and immune challenges have much in common. Stress activates the bidirectional communication of the brain and the immune system. It is also known to activate macrophages for a few days [26]. A human study proved that stress increases Il-1 and Il-6, as well as the cortisol level [27].

Another study on humans showed that acute stress increases IL-1b [28]. These studies confirmed that stress sensitizes microglia and delicately affects the immune system.

The CRF also plays an important role in elevating the pro-inflammatory cytokines level under stress [29]. Most CRF receptors are located in immune cells (i.e., in macrophages, monocytes, and T-lymphocytes). Stress stimulates CRF receptors and increases the level of pro-inflammatory cytokines [29].

6. Tryptophan Metabolism and Pro-inflammatory Cytokines

Pro-inflammatory cytokines increase IDO activity. Increased IDO activation activates the KYN pathway, and in turn, depletes TRP. TRP depletion decreases cell growth and 5-HT synthesis (Figure 2). As mentioned, metabolites of the KYN pathway have both neurotoxic and neuroprotective effects. 3-OH KYN and QUIN have a neurotoxic effect, whereas KYNA has a neuroprotective effect [10]. 3-OH KYN is metabolized to 3-hydroxyanthranilic acid and is auto-oxidated. It results in the production of superoxide anions that have a neurotoxic effect [3]. QUIN has characteristics of the selective N-methyl-D-aspartate (NMDA) agonist, which interrupts the antioxidant defense effect in the cerebral cortex (3,10). On the other hand, KYNA works as an NMDA receptor antagonist and is neuroprotective [3,10]. KYNA also has characteristics of the nicotinic acetylcholine (nACh) receptor agonist, such as the a7-nACh receptor, and is considered neuroprotective [3,10]. A study on macaques (Japanese monkeys) showed that their CSF with Simian-acquired immune deficiency syndrome (SAIDS) showed an increased QUIN/KYNA ratio [30]. A mice study showed that systemic immune activation increased the QUID/KYNA ratio in the CSF [11]. A culture study showed that human microglia and macrophages that were activated using IFN-γ produced KYN and QUIN [31]. This result was interpreted to mean that immunological challenges, including IFN- γ, do increase the production of KYN and QUIN but do not affect KYNA production.

In summary, the balance of the neurotoxic and neuroprotective effects of TRP metabolites is very important in the pathogenesis of neurodegenerative disorders, including of depression.

7. HPA Axis and Pro-inflammatory Cytokines

As mentioned, pro-inflammatory cytokines activate the HPA axis. IL-1 stimulates CRF production, which activates the HPA axis [23]. Pro-inflammatory cytokines allow PGE2 to send signals to BBB [32]. PGE2 production is induced by cyclooxygenase 2 (COX-2) and PGE synthase, and occurs in endothelial cells of cerebral blood vessels and perivascular macrophages. PGE2 activates CRF neurons in PVN. It ends up activating the HPA axis [23]. IFN-γ activates IDO in the hypothalamus and the pituitary gland. Activated IDO increases the production of 2-OH KYN and QUIN, which causes neurodegenerative changes in the HPA axis [33]. Atrophy in the hippocampus is observed in chronic depression, and inhibited negative feedback in the HPA axis is often found, which is attributed to the neurotoxic TRP metabolites.

8. Pro-inflammatory Cytokines and Monoamine

Pro-inflammatory cytokines affect the brain monoamine of each brain region [34]. They activate IDO and the KYN pathway while decreasing the activation of the 5-HT pathway. In a study on mice, the single INF-α injection could not change the monoamine turnover and level, but the repeated INF-α injections decreased the dopamine and 3,4-dihydroxyphenylacetic acid levels [35]. In another mice study, a non-steroidal anti-inflammatory drug (NSAID) was administered to stop the neurochemical response, and acute INF-α injection through an intracerebroventricle increased the 5-HT turnover in the prefrontal cortex and the dopamine turnover in the hippocampus [36]. The extended study used chronic INF-α injection and showed a decreased dopamine level in the prefrontal cortex, and a decreased 5-HT level and 5-HT turnover in the amygdala [37]. In another mice study, continuous INF-α administration consistently increased the 5-HT turnover. However, continuous INF-α administration could not change the mRNA of the 5-HT transporter. [38]. Acute IL-1 injection increased the 5-HT secretion in the CA1 region of the hippocampus but decreased the norepinephrine (NE) in the mice [39]. IL-2 increased the NE turnover in the hypothalamus and hippocampus, and the dopamine turnover in the prefrontal cortex. IL-6 increased the 5-HT level in the prefrontal cortex and hippocampus, as well as the mesocortical dopamine activation [34]. In conclusion, pro-inflammatory cytokines affect the brain monoamine level in various aspects, and its effect may explain the pathogenesis of depression.

9. Pathogenesis of Depression and Pro-Inflammatory Cytokines

Psychological stress or psychiatric disorder is believed to interact with the immune function [40]. The cytokines outside the brain regulate the expression of those in the brain, and vice versa [41]. An animal study confirmed that pro-inflammatory cytokines changed the neuro-endocrine system and caused depression-like behavioral changes [41]. In a human study, an immune therapy using pro-inflammatory cytokines also caused depressive symptoms [42]. Depressed patients are known to have an abnormal immune system, particularly an elevated level of pro-inflammatory cytokines [8]. As such, depression is considered related to dysfunction of the immune and neuroendocrine systems. Pro-inflammatory cytokines elevate CRF production and interrupt negative cortisol feedback (Figure 4). The behavioral changes observed in melancholic depression, including anhedonia or helplessness, also seem to be affected by pro-inflammatory cytokines [41].

In an in-vitro study using human blood, the administration of antidepressant medication using a similar plasma concentration as that in clinical treatment significantly decreased the pro-inflammatory cytokine INF-γ while increasing the immune-inhibiting cytokine IL-10 [43]. In another in-vitro study, the administration of antidepressant medication inhibited the secretion of pro-inflammatory cytokines while increasing the secretion of anti-inflammatory cytokines in human monocytes [44]. An ex-vivo study in depressed patients did not show definite results, which was attributed to the different heterogeneities and clinical conditions of the subjects [45]. However, the result of the in-vitro study on the inhibition, by the antidepressant medication, of the pro-inflammatory cytokine production proved that depression shows immune dysfunction.

5-HT receptors are found in human immunocytes. It is hypothesized that antidepressant medication is associated with the immune mechanism [46]. 5-HT significantly inhibits IFN-γ production, and anti-depressant drugs increase IL-10 secretion [46]. As such, anti-depressant medication decreases the IFN-g/IL-10 ratio. Another mechanism is associated with cAMP. 5-HT stimulates 5-HT receptors in immunocytes and directly activates adenyl cyclase, which increases the intracellular cAMP level. An elevated intracellular cAMP level plays a determinative role in IL-10 and INF-γ production [46]. In a study, the tricyclic anti-depressant (TCA) and the selective serotonin reuptake inhibitor (SSRI) actually elevated the cAMP level in T-lymphocytes and monocytes [47]. However, the anti-depressant effect on pro-inflammatory cytokines has not been clearly proven yet.

10. Monoamine Hypothesis of Depression and Pro-inflammatory Cytokines

As mentioned, pro-inflammatory cytokines are elevated in depression with the changed activity of the brain monoamine. Pro-inflammatory cytokines change the activity of presynaptic NE and presynaptic 5-HT. The regions in the brain of depression patients associated with the aforementioned changes are known to be the prefrontal cortex, hippocampus, hypothalamus, and amygdala [34,36,39]. Many hypotheses on the relationship between pro-inflammatory cytokines and the monoamine system have been suggested. Pro-inflammatory cytokines regulate the activity of the 5-HT transporter [48], and the sensitivity of postsynaptic 5-HT receptors, including of 5-HT_{1A} and 5-HT_{2A} receptors [49]. Pro-inflammatory cytokines also increase COX-2 activity and PGE2 production. Increased PGE2 levels in the CSF and plasma have been reported in depression patients, which support the assumption that pro-inflammatory cytokines contribute to monoamine system dysfunction in depression patients [40]. As such, the activation of pro-inflammatory cytokines and the inflammatory response seem to be associated with the 5-HT system and to cause a depressive system [42].

11. HPA Axis Hyperactivity of Depression and Pro-inflammatory Cytokines

The hyperactivity of the HPA axis and the glucocorticoid resistance of depression patients are already well-known [41]. Pro-inflammatory cytokines are known to activate the CRF system and to play an important role in the pathogenesis of depression. The monocyte-T-lymphocyte hypothesis has been suggested, which explains the determinative role of macrophage activation in the outbreak and pathophysiology of depression [8]. Activated macrophages elevate the IL-1 level, and in turn, elevated IL-1 directly stimulates the secretion of CRF through the PVN neurons in the hypothalamus. According to this hypothesis, cytokines increase the secretion of vasopressin in the parvocellular CRF neurons of PVN under stress and increase the reactivity of the HPA axis. Consistently increasing CRF secretion changes the production and activation of brain cytokines [41]. Pro-inflammatory cytokines activate the HPA axis and increase glucocorticoids. The increased level of glucocorticoids strongly inhibits the immune system.

In summary, stress activates pro-inflammatory cytokines, and the activated pro-inflammatory cytokines cause HPA axis hyperactivity. A persistently hyperactive HPA axis results in immune suppression. This mechanism explains the pathophysiology of the immune system dysfunction observed in depressed patients [9].

12. Tryptophan Metabolism in Depression

The brain TRP level determines the brain 5-HT level. In an animal study, hydrocortisone induced tryptophan 2,3-dioxygenase (TDO) [50], one of the TRP metabolizing enzymes. The TDO induction by hydrocortisone has not yet been confirmed in humans, but hypercortisolemia is known to decrease TRP in animals [51]. As discussed previously, two mechanisms mainly determine the brain TRP level. The first mechanism is the competitive binding of TRP and FFA to plasma albumin. The second mechanism is the competition between TRP and LNAA in BBB (Figure 3). The tyrosine level is closely related to the glucocorticoid level. Glucocorticoid treatment produces a significant amount of hepatic and brain tyrosine aminotransferase, a major tyrosine-degrading enzyme. On the contrary, the tyrosin level controls the glucocorticoid level [52]. Tyrosine competes with LNAA in BBB, and as a result, the glucocorticoid level controls the plasma TRP level.

Glucocorticoids decrease the plasma TRP level and the TRP/LNAA ratio. This result is often observed in depressed patients, whereas healthy subjects show diverse results. A study of healthy subjects showed that dexamethasone decreased the plasma TRP level and the tyrosine level, as well as the TRP/LNAA ratio [53]. However, other studies showed that dexamethasone did not change the plasma TRP level [54,55]. In depressed patients, dexamethasone significantly decreased the TRP level and the plasma TRP/LNAA ratio [53]. Hydrocortisone decreased the plasma TRP in depression patients even after they recovered from depression [55]. The TRP decrease was greater in the patients with major depression than in those with minor depression [56]. In conclusion, the administration of glucocorticoid decreased the TRP level, especially in the depressed patients.

As discussed earlier, cortisol is negatively associated with TRP availability. However, it was confirmed that basal cortisol is not associated with the TRP/LNAA ratio [51, 56]. Another study confirmed that the basal TRP level did not differ between the healthy subjects and the depressed patients even when the LNAA level or the TRP/LNAA ratio was not measured

[57]. In summary, suppressed TRP availability in depressed patients was considered mainly associated with the HPA axis negative feedback dysfunction rather than with the basal glucocorticoid level [51].

Stress and immunological challenges elevate the plasma TRP level and the TRP/LNAA ratio. In contrast, depression and glucocorticoid treatment decrease plasma TRP and the TRP/LNAA ratio. In other words, stress or an immunological challenge in a normal physiological condition changes the peripheral TRP availability through the ß-adrenergic system. On the other hand, the change in the TRP availability occurs through the HPA axis in depression. The change in the TRP availability is determined by the dysfunction of the HPA axis negative feedback, and is not dependent on the basal glucocorticoid level in depression [51]. Under stress, peripheral immune system is activated. Activated immune system increased peripheral IDO and TDO. Increased IDO and TDO activate the TRP consumption. The consumption of peripheral amino acids also increases with the production of acute-phage proteins (APPs) such as haptoglobin and C-reactive protein, as well as TRP [58]. Inflammation decreases plasma APPs, including albumin. As TRP is bound to albumin, albumin directly affects the TRP level. In conclusion, the process of change under stress determines the TRP level and its availability.

Repetitive and persistent exposure to stress is considered a cause of depression. Acute stress increases the plasma TRP level, TRP/LNAA ratio, and brain TRP uptake. However, consistent and repetitive stress leads to physiological adaptation, which eventually breaks down the TRP availability.

A study on depressed patients whose condition was caused by cytokine treatment (i.e., INF-α) showed that the patients with more severe depression symptoms had lower serum TRP levels and higher KYN levels [59]. A study on parous women showed that women with depression and anxiety symptoms during puerperium had increased catabolism from TRP to KYN [60]. The changes in the TRP level and the KYN/TRP ratio were closely associated with the IL-6 level [60].

The depressed patients had elevated pro-inflammatory cytokine levels secreted from the activated macrophages and APPs produced in the liver [40]. They also had decreased plasma TRP levels and TRP/LNAA ratios. The plasma TRP level and IL-6 were negatively correlated [61]. The depression patients also show a decreased albumin level [62].

Many results of studies on the neurotoxic/neuroprotective effects of KYN pathway metabolites on depression have recently been reported. The neurotoxic/neuroprotective activities in the KYN pathway are highlighted as causing depression. Patients on long-term INF-α treatment show increased serum KYN/KYNA ratios. This increased KYN/KYNA ratio indicates a neurotoxic effect. The patients with an increased KYN/KYNA ratio showed a much higher total score in the Montgomery Asberg Depression Rating Scale (MADRS) [12]. Depressed patients are generally known to have lower plasma KYNA levels and higher KYN/KYNA ratios [63]. In the aforementioned study results, both the depressed patients whose condition was induced by INF-α treatment and the depressed patients who experienced no immunological challenge showed decreased KYNA levels. As such, both groups of patients experienced a low neuroprotective effect.

13. Summary of the Tryptophan Metabolism and Pathogenesis of Depression

In this chapter, the TRP metabolism is considered to explain the etiology and pathogenesis of depression. Stress or immunological challenge activates pro-inflammatory cytokines. These activated cytokines activate the peripheral and brain IDO. The activated IDO metabolizes TRP to KYN and activates the KYN pathway, among the two TRP metabolism pathways (the KYN pathway and the 5-HT pathway). Decreased 5-HT pathway and 5-HT production depletes 5-HT, which seems to cause depression. Pro-inflammatory cytokines are involved in presynaptic and postsynaptic 5-HT modulation. These two theories (5-HT depletion and 5-HT modulation by pro-inflammatory cytokines) constitute the monoamine hypothesis of depression.

3-OH KYN and QUIN, the metabolites of the KYN pathways, have a neurotoxic effect, whereas KYNA has a neuroprotective effect. The balance of these two effects seems to play an important role in depression. The imbalance in the metabolites is particularly known to cause neurodegenerative change in the hippocampus.

Pro-inflammatory cytokines activate the HPA axis. This is known as the HPA axis hyperactivity hypothesis of depression. Pro-inflammatory cytokines change the TRP metabolism. The changes in the TRP metabolism induced by pro-inflammatory cytokines, and the immune system activation by stress or immunological challenges, link the mono-amine hypothesis to the HPA axis hyperactivity hypothesis.

1) Imbalance between the KYN pathway induced by IDO and the 5-HT pathway; 2) Neurotoxic/neuroprotective effects of KYN pathway metabolites;

3) Activation of the KYN pathway due to exposure to repeated and consistent stress, which are suggested as the etiology and pathophysiology of depression.

REFERENCES

[1] Hirata, F; Hayaishi, O; Tokuyama, T; Senoh, S. In vitro and in vivo formation of two new metabolites of melatonin. *Journal of Biological Chemistry.*, 1974, 249(4), 1311-3.

[2] Ichiyama, A; Nakamura, S; Nishizuka, Y; Hayaishi, O. Enzymic studies on the biosynthesis of serotonin in mammalian brain. *Journal of Biological Chemistry.*, 1970, 245(7), 1699-709.

[3] Stone, TW; Darlington, LG. Endogenous kynurenines as targets for drug discovery and development. *Nature Reviews Drug Discovery.*, 2002, 1(8), 609-20.

[4] Widner, B; Laich, A; Sperner-Unterweger, B; Ledochowski, M; Fuchs, D. Neopterin production; tryptophan degradation; and mental depression—What is the link? *Brain, behavior; and immunity.* 2002, 16(5), 590-5.

[5] Schruers, K; Griez, E. The effects of tryptophan depletion on mood and psychiatric symptoms. *Journal of affective disorders.*, 2003, 74(2), 203.

[6] Hoes, M; Sijben, N. The clinical significance of disordered renal excretion of xanthurenic acid in depressive patients. *Psychopharmacology.*, 1981, 75(4), 346-9.

[7] Schüle, C. Neuroendocrinological mechanisms of actions of antidepressant drugs. *Journal of neuroendocrinology.*, 2007, 19(3), 213-26.

[8] Smith, R. The macrophage theory of depression. *Medical hypotheses.*, 1991, 35(4), 298-306.

[9] Maes, M; Smith, R; Simon, S. The monocyte-T-lymphocyte hypothesis of major depression. *Psychoneuroendocrinology.*, 1995, 20(2), 111-6.

[10] Myint, AM; Kim, YK. Cytokine–serotonin interaction through IDO, a neurodegeneration hypothesis of depression. *Medical hypotheses.*, 2003, 61(5), 519-25.

[11] Saito, K; Crowley, J; Markey, S; Heyes, M. A mechanism for increased quinolinic acid formation following acute systemic immune stimulation. *Journal of Biological Chemistry.*, 1993, 268(21), 15496-503.

[12] Wichers, M; Koek, G; Robaeys, G; Verkerk, R; Scharpe, S; Maes, M. IDO and interferon-α-induced depressive symptoms, a shift in

hypothesis from tryptophan depletion to neurotoxicity. *Molecular psychiatry.*, 2005, 10(6), 538-44.

[13] Dunn, AJ. Systematic interleukin-1 administration stimulates hypothalamic norepinephrine metabolism parallelling the increased plasma corticosterone. *Life sciences.*, 1988, 43(5), 429-35.

[14] Kabiersch, A; del Rey, A; Honegger, CG; Besedovsky, HO. Interleukin-1 induces changes in norepinephrine metabolism in the rat brain. *Brain; behavior; and immunity.*, 1988, 2(3), 267-74.

[15] Dunn, AJ; Welch, J. Stress-and endotoxin-induced increases in brain tryptophan and serotonin metabolism depend on sympathetic nervous system activity. *Journal of neurochemistry.*, 1991, 57(5), 1615-22.

[16] Chaouloff, F. Effects of acute physical exercise on central serotonergic systems. *Medicine & Science in Sports & Exercise.*, 1997.

[17] Strüder, H; Hollmann, W; Platen, P; Wöstmann, R; Weicker, H; Molderings, G. Effect of acute and chronic exercise on plasma amino acids and prolactin concentrations and on [3H] ketanserin binding to serotonin2A receptors on human platelets. *European journal of applied physiology and occupational physiology.*, 1999, 79(4), 318-24.

[18] Guillemin, GJ; Smythe, G; Takikawa, O; Brew, BJ. Expression of indoleamine 2; 3-dioxygenase and production of quinolinic acid by human microglia, astrocytes, and neurons. *Glia.*, 2005, 49(1), 15-23.

[19] Roy, EJ; Takikawa, O; Kranz, DM; Brown, AR; Thomas, DL. Neuronal localization of indoleamine 2, 3-dioxygenase in mice. *Neuroscience letters.*, 2005, 387(2), 95-9.

[20] Pawlak, D; Takada, Y; Urano, T; Takada, A. Serotonergic and kynurenic pathways in rats exposed to foot shock. *Brain research bulletin.* 2000, 52(3), 197-205.

[21] Miura, H; Ozaki, N; Shirokawa, T; Isobe, K. Changes in brain tryptophan metabolism elicited by ageing, social environment, and psychological stress in mice. Stress, *The International Journal on the Biology of Stress.*, 2008, 11(2), 160-9.

[22] Maes, M; Song, C; Lin, A; De Jongh, R; Van Gastel, A; Kenis, G; et al. The effects of psychological stress on humans, increased production of pro-inflammatory cytokines and Th1-like response in stress-induced anxiety. *Cytokine.*, 1998, 10(4), 313-8.

[23] Black, PH. Immune system-central nervous system interactions, effect and immunomodulatory consequences of immune system mediators on the brain. *Antimicrobial Agents and Chemotherapy.*, 1994, 38(1), 7-12.

[24] Maier, SF. Bi-directional immune–brain communication, implications for understanding stress, pain, and cognition. *Brain, behavior, and immunity.*, 2003, 17(2), 69-85.
[25] Banks, W; Kastin, A; Broadwell, R. Passage of cytokines across the blood-brain barrier. *Neuroimmunomodulation.*, 1995, 2(4), 241-8.
[26] LeMay, LG; Vander, AJ; Kluger, MJ. The effects of psychological stress on plasma interleukin-6 activity in rats. *Physiology & behavior.*, 1990, 47(5), 957-61.
[27] Schulte, H; Bamberger, C; Elsen, H; Herrmann, G; Bamberger, A; Barth, J. Systemic interleukin-1 α and interleukin-2 secretion in response to acute stress and to corticotropin-releasing hormone in humans*. *European journal of clinical investigation.*, 1994, 24(11), 773-7.
[28] Minami, M; Kuraishi, Y; Yamaguchi, T; Nakai, S; Hirai, Y; Satoh, M. Immobilization stress induces interleukin-1β mRNA in the rat hypothalamus. *Neuroscience letters.*, 1991, 123(2), 254-6.
[29] Leonard, BE. The HPA and immune axes in stress, the involvement of the serotonergic system. *European Psychiatry.*, 2005, 20, S302-S6.
[30] Heyes, MP; Mefford, IN; Quearry, BJ; Dedhia, M; Lackner, A. Increased ratio of quinolinic acid to kynurenic acid in cerebrospinal fluid of D retrovirus—infected rhesus macaques, Relationship to clinical and viral status. *Annals of neurology.*, 1990, 27(6), 666-75.
[31] Heyes, M; Achim, C; Wiley, C; Major, E; Saito, K; Markey, S. Human microglia convert l-tryptophan into the neurotoxin quinolinic acid. *Biochem J.*, 1996, 320, 595-7.
[32] Konsman, JP; Parnet, P; Dantzer, R. Cytokine-induced sickness behaviour, mechanisms and implications. *Trends in neurosciences.*, 2002, 25(3), 154-9.
[33] Tu, H; Rady, PL; Juelich, T; Smith, EM; Tyring, SK; Hughes, TK. Cytokine regulation of tryptophan metabolism in the hypothalamic-pituitary-adrenal (HPA) axis, implications for protective and toxic consequences in neuroendocrine regulation. *Cellular and molecular neurobiology.*, 2005, 25(3-4), 673-80.
[34] Zalcman, S; Green-Johnson, JM; Murray, L; Nance, DM; Dyck, D; Anisman, H; et al. Cytokine-specific central monoamine alterations induced by interleukin-1;-2 and-6. Brain research. 1994, 643(1), 40-9.
[35] Shuto, H; Kataoka, Y; Horikawa, T; Fujihara, N; Oishi, R. Repeated interferon-α administration inhibits dopaminergic neural activity in the mouse brain. *Brain research.*, 1997, 747(2), 348-51.

[36] De La Garza, R; Asnis, GM. The non-steroidal anti-inflammatory drug diclofenac sodium attenuates IFN-α induced alterations to monoamine turnover in prefrontal cortex and hippocampus. *Brain research.*, 2003, 977(1), 70-9.

[37] De La Garza, R; Asnis, GM; Pedrosa, E; Stearns, C; Migdal, AL; Reinus, JF; et al. Recombinant human interferon-α does not alter reward behavior, or neuroimmune and neuroendocrine activation in rats. *Progress in Neuro-Psychopharmacology and Biological Psychiatry.* 2005, 29(5), 781-92.

[38] Sato, T; Suzuki, E; Yokoyama, M; SEMBA, JI; Watanabe, S; Miyaoka, H. Chronic intraperitoneal injection of interferon-α reduces serotonin levels in various regions of rat brain, but does not change levels of serotonin transporter mRNA, nitrite or nitrate. *Psychiatry and clinical neurosciences.*, 2006, 60(4), 499-506.

[39] Broderick, PA. Interleukin 1α alters hippocampal serotonin and norepinephrine release during open-field behavior in Sprague–Dawley animals, differences from the Fawn-Hooded animal model of depression. *Progress in Neuro-Psychopharmacology and Biological Psychiatry.*, 2002, 26(7), 1355-72.

[40] Leonard, BE. Changes in the immune system in depression and dementia, causal or co-incidental effects? *International Journal of Developmental Neuroscience.*, 2001, 19(3), 305-12.

[41] Dantzer, R; Wollman, E; Vitkovic, L; Yirmiya, R. Cytokines and depression, fortuitous or causative association? *Molecular Psychiatry.*, 1999, 4(4), 328-32.

[42] Bonaccorso, S; Meltzer, H; Maes, M. Psychological and behavioural effects of interferons. Current Opinion in Psychiatry. 2000, 13(6), 673-7.

[43] Song, C. The interaction between cytokines and neurotransmitters in depression and stress, possible mechanism of antidepressant treatments. Human Psychopharmacology, *Clinical and Experimental.*, 2000, 15(3), 199-211.

[44] Kubera, M; Kenis, G; Bosmans, E; Scharpé, S; Maes, M. Effects of serotonin and serotonergic agonists and antagonists on the production of interferon-γ and interleukin-10. *Neuropsychopharmacology.*, 2000, 23(1), 89-98.

[45] Haack, M; Hinze-Selch, D; Fenzel, T; Kraus, T; Kühn, M; Schuld, A; et al. Plasma levels of cytokines and soluble cytokine receptors in psychiatric patients upon hospital admission, effects of confounding

factors and diagnosis. *Journal of psychiatric research.*, 1999, 33(5), 407-18.

[46] Maes, M. The immunoregulatory effects of antidepressants. Human Psychopharmacology, *Clinical and Experimental.*, 2001, 16(1), 95-103.

[47] Xia, Z; DePierre, J; Nassberger, L. Tricyclic antidepressants inhibit Il-6, IL-1 beta, and TNF alpha release in human blood monocytes and IL-2 and IFN gamma in T cells. *Immunobiology.*, 1996, 34, 37-47.

[48] Ramamoorthy, S; Ramamoorthy, JD; Prasad, PD; Bhat, GK; Mahesh, VB; Leibach, FH; et al. Regulation of the human serotonin transporter by interleukin-1β. *Biochemical and biophysical research communications.*, 1995, 216(2), 560-7.

[49] Abe, S; Hori, T; Suzuki, T; Baba, A; Shiraishi, H; Yamamoto, T. Effects of chronic administration of interferon α A/D on serotonergic receptors in rat brain. *Neurochemical research.*, 1999, 24(3), 359-63.

[50] Hayaishi, O; Rothberg, S; Mehler, AH; Saito, Y. STUDIES ON OXYGENASES ENZYMATIC FORMATION OF KYNURENINE FROM TRYPTOPHAN. *Journal of Biological Chemistry.*, 1957, 229(2), 889-96.

[51] Porter, RJ; Gallagher, P; Watson, S; Young, AH. Corticosteroid-serotonin interactions in depression, a review of the human evidence. *Psychopharmacology.*, 2004, 173(1-2), 1-17.

[52] HIROTA, T; HIROTA, K; SANNO, Y; TANAKA, T. A New Glucocorticoid Receptor Species, Relation to Induction of Tryptophan Dioxygenase by Glucocorticoids*. *Endocrinology.*, 1985, 117(5), 1788-95.

[53] Maes, M; Jacobs, MP; Suy, E; Minner, B; Leclercq, C; Christiaens, F; et al. Suppressant effects of dexamethasone on the availability of plasma L-tryptophan and tyrosine in healthy controls and in depressed patients. *Acta Psychiatrica Scandinavica.*, 1990, 81(1), 19-23.

[54] Porter, RJ; McAllister-Williams, RH; Jones, S; Young, AH. Effects of dexamethasone on neuroendocrine and psychological responses to L-tryptophan infusion. *Psychopharmacology.*, 1999, 143(1), 64-71.

[55] Bhagwagar, Z; Hafizi, S; Cowen, P. Cortisol modulation of 5-HT-mediated growth hormone release in recovered depressed patients. *Journal of affective disorders.*, 2002, 72(3), 249-55.

[56] Maes, M; Schotte, C; Scharpe, S; Martin, M; Blockx, P. The effects of glucocorticoids on the availability of L-tryptophan and tyrosine in the plasma of depressed patients. *Journal of affective disorders.*, 1990, 18(2), 121-7.

[57] Cowen, PJ; Charig, EM. Neuroendocrine responses to intravenous tryptophan in major depression. *Archives of general psychiatry.*, 1987, 44(11), 958-66.

[58] Schiepers, OJ; Wichers, MC; Maes, M. Cytokines and major depression. *Progress in Neuro-Psychopharmacology and Biological Psychiatry.*, 2005, 29(2), 201-17.

[59] Bonaccorso, S; Marino, V; Puzella, A; Pasquini, M; Biondi, M; Artini, M; et al. Increased depressive ratings in patients with hepatitis C receiving interferon-α–based immunotherapy are related to interferon-α–induced changes in the serotonergic system. *Journal of clinical Psychopharmacology.*, 2002, 22(1), 86-90.

[60] Maes, M; Verkerk, R; Bonaccorso, S; Ombelet, W; Bosmans, E; Scharpé, S. Depressive and anxiety symptoms in the early puerperium are related to increased degradation of tryptophan into kynurenine, a phenomenon which is related to immune activation. *Life sciences.*, 2002, 71(16), 1837-48.

[61] Maes, M; Meltzer, HY; Scharpè, S; Bosmans, E; Suy, E; De Meester, I; et al. Relationships between lower plasma L-tryptophan levels and immune-inflammatory variables in depression. *Psychiatry research.*, 1993, 49(2), 151-65.

[62] Maes, M; Wauters, A; Verkerk, R; Demedts, P; Neels, H; Van Gastel, A; et al. Lower serum L-tryptophan availability in depression as a marker of a more generalized disorder in protein metabolism. *Neuropsychopharmacology.*, 1996, 15(3), 243-51.

[63] Myint, A-M; Kim, YK; Verkerk, R; Scharpé, S; Steinbusch, H; Leonard, B. Kynurenine pathway in major depression, evidence of impaired neuroprotection. *Journal of affective disorders.*, 2007, 98(1), 143-51.

In: New Developments in Tryptophan Research ISBN: 978-1-63483-025-6
Editor: Victoria Hayes

Chapter 3

L-TRYPTOPHAN IN PERINATAL NUTRITION, A PUTATIVE ROLE IN CIRCADIAN RHYTHM ACQUISITION?

Elizabeth Nascimento* and Bertrand Kaeffer
E Nascimento, Departamento de Nutrição, Centro de Ciências da Saude, Universidade Federal de Pernambuco, Recife, Pernambuco, Brazil,
B Kaeffer, Unité Mixte de Recherche-1280, Physiologie des Adaptations Nutritionnelles, Institut National Recherche Agronomique, Université de Nantes, France

ABSTRACT

Mammalian cells cannot synthesize the amino acid tryptophan and dietary intake of tryptophan is crucial in condition of rapid tissue growth and catabolic cellular stress. These situations occur in early life, especially with babies who are born underweight. Tryptophan is metabolized in peripheral tissues such as gastrointestinal tract and liver, as well as by the central nervous system. This amino acid is a precursor of key molecules of circadian cycles and energy homeostasis such as serotonin, melatonin and niacins depending on its time of administration. This review will illustrate the key role of L-tryptophan in metabolic functions with special emphasis on the alterations of circadian rhythms.

* Email: nlizbeth@gmail.com or Bertrand.Kaeffer@univ-nantes.fr

INTRODUCTION

The tryptophan amino acid was discovered by Hopkins and Cole (Hopkin & Cole, 1901) in the early 1900s after isolating it from casein protein. Ellinger and Flamand (1908) determined its molecular structure a short time later. Only isoform L is used in organisms. The tryptophan was first synthesized in 1949; but the early 1980's chemical synthesis of tryptophan was replaced with fermentation procedures that greatly increased obtainable yields; making tryptophan supplements more available (Sidransky, 2002). Tryptophan is metabolized in peripheral tissues such as gastrointestinal tract and liver, as well as by the central nervous system (Geeraerts et al., 2011, Bertrand & Bertrand, 2010). Among the amino acids forming proteins, tryptophan is in proportion amongst the lowest.

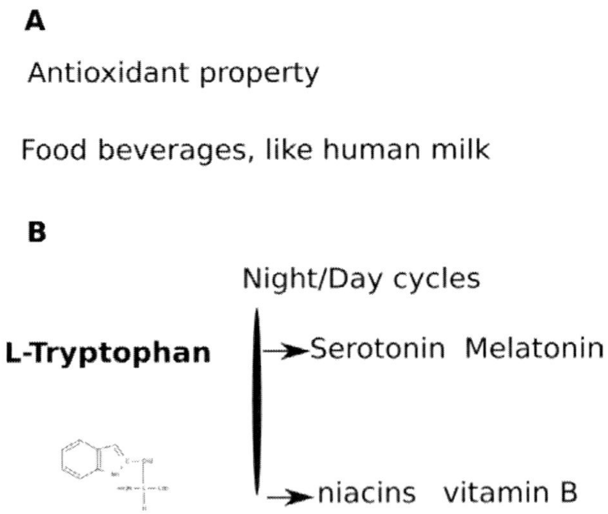

Figure 1. L-Tryptophan, an essential amino acid with pleiotropic effects.Besides its antioxidant properties in food beverage like breast milk (A), L-Tryptophan has pleiotropic effects on general physiology (biosynthesis of protein, neurotransmitter serotonin, hormone melatonin and of niacin). Therefore, the biosynthesis of derivatives or degradation products from tryptophan (B) can depend on Night/Day cycles, on the protein diet, tryptophan/large neutral amino acids (LNAA) ratio and others components such as carbohydrates and lipids (Fernstrom et al., 1979) these nutrients also influence uptake of tryptophan by cell expressing L-type amino acid transporter transporters (Kaper et al., 2007). The insulin and TRP/LNAA ratio is important for uptake and synthesis of brain serotonin.

Tryptophan from protein as well as free and protein-bound tryptophan have been the subject to extensive research (Bertazzo et al., 2011). Dietary intake of tryptophan is crucial in condition of rapid tissue growth (Kaper et al., 2007) and catabolic cellular stress, a situation that occurs in children in early life, especially those who are born underweight (Chessex et al., 2010). Protein or energy restriction can modify availability and metabolism of tryptophan influencing rhythms of circadian clocks and therefore food patterns. L-tryptophan has pleiotropic effects on general physiology (biosynthesis of protein, of the neurotransmitter serotonin, the hormone melatonin and of niacins; Figure 1). The degradation products of tryptophan have immunoregulatory, antioxidants functions (Tsopmo et al., 2009; Chessex et al., 2010) and some tryptophan photoproducts have direct effects on aryl hydrocarbon receptors of skin cells, i.e., on skin circadian clock (Mukai & Tischkau, 2007). This review will illustrate the key role of L-tryptophan in metabolic functions with special emphasis on the alteration of circadian rhythms.

1. L-Tryptophan Role in Mammalian Metabolism Depends on the Time of Consumption and Interactions with other Nutrients

L-tryptophan is usually considered as the less abundant amino acid in the diet (Fazzolari-Nesci et al., 1992; Richard et al., 2009). In humans, tryptophan has relatively low tissue storage and the overall concentration in the body is the lowest among all amino acids (Richard et al., 2009). In humans the daily need is estimated between 3.5 to 6 mg L-tryptophan /kg/ of body weight per day (Richard et al., 2009). In rats for an optimal growth, the estimated value is of 2.0g/kg of diet (Benevenga et al., 1994). Because tryptophan is found in the lowest range of concentrations among amino acids; it is relatively less available and is thought to play a rate-limiting role during protein synthesis (Sainio et al., 1996) or in metabolic pathways of kynurenine (Chessex et al., 2010), niacin and derivated coenzymes (Chessex et al., 2010; Geeraerts et al., 2011), as well as for the synthesis of melatonin (Sainio et al., 1996) and serotonin (Sainio et al., 1996).

Dopamine, norepinephrine and beta-endorphin have been shown to increase following oral dose of tryptophan (Sanger, 2008; Maes et al., 1989). The amino acid is also thought to be involved in the turn-over of cortisol and noradrenaline (Maes et al., 1989) as well as on the plasma levels of prolactin and growth hormone (Kollmann et al., 2008).

1.1. Biosynthesis of Serotonin

L-tryptophan is converted in 5-hydroxy-tryptophan (5-HTP) by hydroxylases and then the 5-HTP is converted in 5-hydroxy-Tryptamin (5-HT) or serotonin by decarboxylases. Tryptophan hydroxylases in human (Meyer & Brinck, 1999) have been detected not only in the central nervous system but also in normal enterocytes and in some cells of secretory tubules of the pyloric mucosa or of the duodenal glands of Brunner suggesting that epithelial cells of the small intestine (and some cells of the stomach or duodenum) are able to synthetize 5-hydroxytryptophan (Bertrand & Bertrand, 2010). The activities of tryptophan hydroxylases (TPH1 and TPH2) in the gastrointestinal tract are different from brain and relatively little is known about the factors controlling peripheral serotonin synthesis (Colmenares et al., 1975**).** The TPH1 isoforms are expressed mainly in stomach, small intestine and enterochromaffin cells (Kanai & Endou, 2003).

The TPH2 isoform is expressed mainly in brain (Kanai & Endou, 2003), but this isoform, has been localized also in central raphe neurons and in enteric neurons (Kanai & Endou, 2003; Malek et al., 2005). Tryptophan and other large neutral amino acids share a common carrier (LAT- large amino acid transporter), but the affinity of this transport system for each amino acid is at least an order of magnitude lower than of the blood brain barrier transport system, e g, the Km of gut amino acid transport is at least tenfold greater than normal levels of plasma amino acids (Colmenares et al., 1975; Figure 2). It is estimated that only 1% of dietary tryptophan is used for serotonin synthesis in the brain (Baker, 2005).

The immediate decarboxylation of tryptophan results in the synthesis of trace amounts of tryptamine (i.e., ng/g), which is an important neuromodulator of serotonin (Jones, 1982). Some animal studies have indicated that tryptamine acts as a control for the balance between excitatory and inhibitory functions of serotonin, and in other instances, tryptamine acts as a neurotransmitter with specific receptors that are independent of serotonin function (Jones, 1982). Therefore, the biosynthesis of metabolites or the degradation products of tryptophan can depend on the proteins diet, tryptophan/large neutral amino acids (LNAA) ratio and others components such as carbohydrates and lipids (Fernstrom & Wurtman, 1979), whose nutrients also influence uptake of tryptophan by cell expressing L-type amino acid transporter (Kaper et al., 2007). The insulin and TRP/LNAA ratio are important for uptake and synthesis of brain serotonin, but not in the gastrointestinal tract (Malek et al., 2005; Penkova et al., 2010; Yu et al., 1999). In the brain the serotonin is a substrate of melatonin synthesis, one indole is important in other neural process like circadian rhythms and sleep regulation.

The total amount of serotonin in human organism is about 10 mg; 95% are found in gastrointestinal tract among which 90% are in enterocromafins cells (epithelium and glands) and 10% in the mast cells from the wall of the digestive tube. Around 95% or more of the whole body serotonin is produced by the gastrointestinal tract, however this peripheral production cannot be utilized by the central nervous system because the molecule cannot cross the blood brain barrier (Bertrand & Bertrand, 2010; Malek et al., 2005) except when the blood brain barrier is not formed (Coté et al., 2006). Study utilizing mouse genetically altered for isoform TPH1 have shown that maternal serotonin is crucial for the development of the central nervous system during pregnancy. In mouse the blood brain barrier is formed during post-natal life and alterations of serotonin from dams influence development of the nervous system of pups (Coté et al., 2006).

The serotonin molecule is implicated in the normal ontogenesis of the nervous system (Deiro et al., 2008), in food intake (Lopes De Souza et al., 2008), or in feeding behavior (Leibowitz & Alexandre, 1998), aggressive behavior (Manhaes De Castro et al., 2001), body temperature , sleeps control (Steinberg et al., 2011), and gut functions with intestinal secretion and motility (Bertrand & Bertrand, 2010; Penkova et al., 2010; Meyer & Brinck, 1989). Recent works demonstrate the implications of serotonin in inflammatory disease and irritable intestine syndrome (Penkova et al., 2010).

The capacity of enteroendocrine cells, mast cells and nerve cell bodies and fibers in the gastrointestinal tracts of both human and rat to synthesize serotonin from tryptophan has been explored by immunohistochemistry (Yu et al., 1999).

The roles of serotonin and tryptophan in the metabolism of gastrointestinal tract have been less explored than in the central nervous system. Of the additional substances contained in enterochromaffin (EC) cells of intestinal tract is melatonin (Bertrand & Bertrand, 2010). The exploration of this synthesis mechanism is little explored partly because of difficulty due to *in vitro* cell cultures and due to the relatively small proportion of the EC cells, perhaps, 1-3% in human intestinal cells (Coates et al., 2004). However the role of serotonin in gastrointestinal tract has been described by several authors.

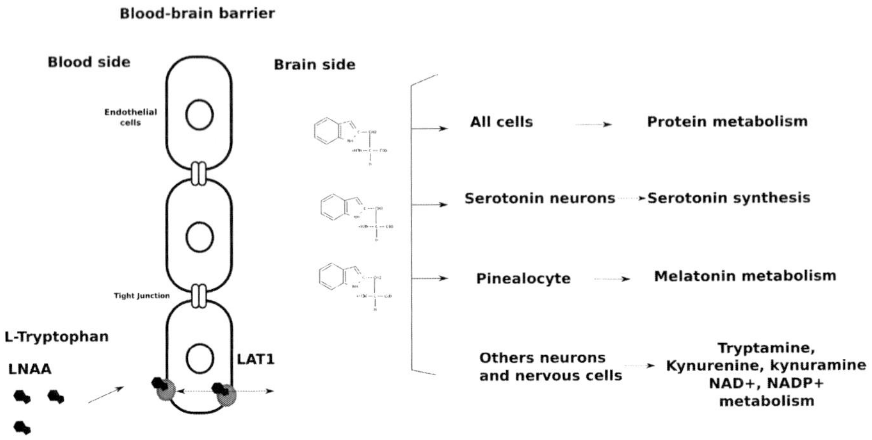

Figure 2. Metabolic pathway of tryptophan in the central nervous system. Free tryptophan is transported across the blood-brain barrier via the L-type amino acid transporter (LAT1)/4f2 heavy chain complex at the luminal and abluminal surfac es of endothelial cells. Once in the cerebrospinal fluid, tryptophan is available for uptake into all cells for protein synthesis and specialised cells that synthesise neuroactive metabolites of tryptophan, including serotonin, tryptamine, kynurenines. In the brain the serotonin is substrate of melatonin synthesis, one indole important of other neural process circadian rhythms and sleep regulation (Ruddick et al., 2006).

The serotonin released from EC cells participate to gastrointestinal motility (Geeraerts et al., 2011; Bertrand & Bertrand, 2010) , rhythms of digestion (Geeraerts et al., 2011; Bertrand & Bertrand, 2010), intestinal disease (Gershon & Tack, 2007), maturation of gastrointestinal tract (Bertrand & Bertrand, 2010) as well as implications in inflammatory diseases and syndrome of the irritable intestine (Penkova et al., 2010).

The main site of peripheral serotonin synthesis is the EC cells of the gut which function as sensory transducers using serotonin as paracrine messenger (Gershon & Tack, 2007). The serotonin transporter in the native human intestine has been studied only recently at the duodenum, jejunum and ileum levels providing evidence that a functional serotonin transporter exists and is capable of removing intraluminal serotonin in human ileal epithelial cells (Gill et al., 2008). The cellular activity is also modulated by some nutrients (Bertrand & Bertrand, 2010).

The modulatory pathway between tryptophan, serotonin and metabolites in gut is complex. Thus, association of these biological processes with the central and peripheral oscillators of circadian rhythms and the connection between these mechanisms are even less known.

1.2. Serotonin and Melatonin Regulate the Circadian Clock System, Effects of L-Tryptophan Intake on this Regulation

Mammals exhibit behavioral and physiological rhythms which are driven by an endogenous timing system called the circadian clock. This system is entrained by light-dark cycles and other environmental stimuli such as food (Nakahata et al., 2007) and several signaling molecules including serotonin (*Figure 1B*). In the brain, circadian rhythm of serotonin receptors have been described in humans (Reinberg, 2003) and non-photic stimuli, such as food shortage or serotonergic receptor activation by 5-HT input from Raphe nuclei are also capable of shifting the master clock and/or modulating photic synchronization (Nakahata et al., 2007).

Fuller et al., 2008 have shown that one food-entrainable clock may be located in the dorsomedial nucleus of the hypothalamus. According to Green & Menaker (2003), the molecular loops that generate circadian rhythms may display distinct molecular complexes related to clock genes, one in the forebrain and another in suprachiasmatic nuclei with different synchronizers. The serotoninergic neurons reaches the region of the suprachiasmatic nucleus constituting the input of the central oscillator (Fuller et al., 2008). However circadian clocks reside not only in suprachiasmatic nuclei but in other brain regions and peripheral organs and even in cultured cells (Yu et al 1999).

The indole melatonin is considered to be the principal component of the internal circadian clock located in the suprachiasmatic nucleus (SNC). The melatonin production is subjected to circadian rhythms with high levels at night and low levels by day, and also to seasonal variations throughout the year (Kennaway & Wright, 2002).

In the central nervous system, the biosynthesis of melatonin is dependent on serotonin availability (Kennaway & Wright, 2002) and on the tryptophan metabolic pathway. The serotonin is converted into melatonin by the enzyme aralkylamine N-actyltransferase (AA-NAT), which constitutes the limiting step in melatonin synthesis. This enzyme presents a marked circadian rhythm in all species that have been studied (Arendt, 2006). However, the final step in the biosynthesis of melatonin is catalysed by hydroxyindole-o-methyltransferase (HIOMT), which does not show circadian activity (Klein, 1997).

The melatonin is primarily synthesized by pinealocytes of pineal gland during the night regardless of the diurnal or nocturnal locomotor acitivity (Liu & Borjigin, 2010; Chowdhuri, 2008). The pineal gland of all vertebrates produces melatonin from tryptophan, which is converted into serotonin in response to the norepinephrine released from the superior cervical ganglion (Borjigin et al., 1999). The magnitude and duration of nocturnal synthesis is dependent upon the length of dark phase and by the interacting network of circadian clock genes of SNC. The melatonin is taken up by all tissues, but its catabolic mechanism is less understood (Chowdhuri, 2008). The chronobiotic effects of melatonin is caused by its direct influence on the electrical and metabolic activity of SNC according to *in vitro* and *in vivo* experiments (Pévet et al., 2006; Kennaway & Wright, 2002; Nakahata et al., 2007).

The balance between the use of serotonin (5-HT) and its availability as precursor is determinant for biosynthesis of melatonin. Both serotonin and melatonin are fundamental for the maintenance of the functionality of the central nervous system, and for the regulation of the circadian clock located in the suprachiasmatic nucleus (Green & Menaker, 2003; Reinberg, 2003; Arendt, 2006).

The importance of synchronization of circadian rhythm for brain development is manifest because the neuronal activation occurs mainly during sleep (Krueger et al., 2008). The period of sleep of foetus and babies, for 16 to 18 hours per day is important to brain growth (Ednick et al., 2009). Alteration in neurogenesis of rat had been observed. Lactating neonatal rats fed with different formulas containing several combinations of nutrients, such as highly disponible tryptophan, may facilitate the circadian wake-sleep rhythm (Puigjaner et al., 2007). Periodic breast feeding and maternal care entrain molecular oscillations in the peripheral clocks. Although the SCN clock of newborn rat pups is already photosensitive, the photic entrainment does not yet occur until P6-P20 (Jin et al., 1999). During this period, pups open their eyes and start to be partly independent of their mothers (Sumová et al., 2006). This last information explains, why we had chosen "light off" as the main temporal time point to apply our tryptophan bolus in our experiment of supplementation of rat pups by a bolus of tryptophan (Nascimento et al., 2013). The molecular machinery which drives circadian oscillations is maturing at different times in different tissues and the phase of some peripheral organs is malleable and may shift as the organ's function changes during development (Yamazaki et al., 2009). Activation of the serotoninergic pathway from the raphe nuclei (which release serotonin) acts as a non-photic synchronizer of the SNC (Meyer-Bernstein & Morin, 1996). Periodic breast feeding and maternal care entrain molecular oscillations of the peripheral clocks. Expression of clock genes in gastric epithelial cells has been also described in preterm infants (Kaeffer et al., 2011).

In rat, apart from the maternal day-time feeding, pups begin gradually to forage themselves during the night and molecular oscillations of peripheral clocks shift accordingly between P20–P28. During these few days, rat pups become completely independent of their mothers and the maternal entrainment is lost. Similarly as in adults, the SCN clock is entrained dominantly by photic cues and peripheral clocks by nocturnal feeding regime. Maternal melatonin may fulfil all the criteria of a functional entrainment cue and is considered as a first class candidate for the photoperiodic entrainment (Sumovà et al., 2006).

The development of peripheral clocks depends on maturation of the organ and the first appearance of molecular oscillations may be highly dependent on organs and species (Sumová et al., 2003). The synchronization of this rhythm also influences the development of preadipocytes and myocytes (Bray & Young, 2006; Young, 2006).

In the murine cerebral cortex, daily rhythms of Per1 and Per2 mRNA are detected from P14 (Shimomura et al., 2001). The foetal clock is supposed to be entrained exclusively by cues delivered periodically from the mother because the photic pathways to the foetal SCN in altricial rodents are not fully developed (Sumová et al., 2003; Sumovà et al., 2006). First, the rhythm in the foetal SCN metabolic activity is synchronized by the maternal SCN (Coté et al., 2006). Second, these new rhythms of clock genes in the very late foetal and early neonatal stages are, from the beginning, in phase with the maternal clock (Sladek et al., 2006). In rats, it is suggested that the maternal synchronization of foetal clocks occurs even before the SCN is formed. There is also confusion concerning the photoperiodic entrainment of foetuses and newborn rodents. The work of Varcoe et al (2013) confirms and enhances these observations. Starting from prenatal stage at E20, a newborn completely dependent on maternal care progresses gradually from of stage where light starts to entrain the SCN clock at P10 up to the time of weaning in P20, until the pup reaches a mature state around P30 and becomes gradually completely independent of maternal care (Sumová et al., 2003). In liver, the expression of clock genes (per1, per2, rev-erba; cry1, bmal1 clock) are circadian and the pre and postnatal development of the molecular clockwork in the rat liver proceeds gradually with rev-erb-a exhibiting a significant rhythm throughout all the studied developmental stages (i.e., E20 until adulthood). Oscillationsof high amplitude of all other clock genes are detectable only at P10 and later (Sumová et al., 2003; Sumovà et al., 2006). In the cerebral cortex of mouse, daily rhythms of Per1 and Per2 mRNA are detected from P14 (Shimomura et al., 2001). The foetal rhythm of the metabolic activity in SCN is synchronized by maternal SCN (Coté et al., 2006). In the rat heart, circadian expression of clock genes PER1 and BMAL1 starts between P2 and P5, while PER2 expression is not rhythmic until P14 (Penedo et al., 2009) and adult patterns of heart clock gene expression are only encountered after P20. Thus for peripheral tissues as well as for SCN, the rhythmic expression of clock genes appears gradually during early post natal development. In mouse liver, many clock-controlled genes encode enzymes involved in food processing (e.g., phosphoenolpyruvate carboxykinase, glycogen synthase), heme synthesis (aminolevulinate synthase), detoxification (components of cytochrome P450), bile acids synthesis (cholesterol 7 a-hydroxylase) (Panda et al., 2002).

Both the circadian expression of genes and the synchronization of clocks are influenced by several factors including feed profile and diet composition (Mendoza et al., 2008) and molecules like serotonin and its non-photic agonists (Meyer-Bernstein & Morin, 1996). Also, activation of several brain functions are dependent on the levels of amino acids or metabolites that act like neurotransmitter (Penedo et al., 2009) or signal regulatory of cell activation such as GABAergic system (Feria-Velasco et al., 2008).

2. L-Tryptophan As a Substrate for Synthesis of the Coenzymes Nicotinamide Adenine Dinucleotide and Its Phosphorylated Form

Tryptophan also plays a role as a substrate for synthesis of the coenzymes nicotinamide adenine dinucleotide (NADH) and its phosphorylated form, NAD-phosphate (NADPH). These coenzymes influence the affinity of CLOCK/BMAL1 and NPAS2/BMAL1 complexes for their target DNA (Zhang et al., 2009). NAD+ and NADP are coenzymes essential for electron transfer reactions (i.e., redox reactions) and NAD+ levels in all mammalian cells is crucial metabolites of modulation of SIRT1 activity, an histone protein deacetylase, that modulates activity of CLOCK:BMAL1 dimers involved in the expression of clock-driven genes (Li & Lin, 2009; Zhang et al, 2009). These coenzymes can be synthesized not only from ingested tryptophan, but also by ingestion of niacin (i.e., vitamin B3). On the other hand, the NAD+ synthesis is dependent on Nicotinamide phosphoribosyltransferase (NAMPT) expression and activity. This rate-limiting enzyme converts nicotinamide to nicotinamide mononucleotide in the NAD biosynthetic pathway in mammals (Revollo et al., 2007; Zhang et al., 2009).

Tryptophan metabolites like kynurenines have been related with the dietary intake of tryptophan and niacin with consequences on the mood of healthy men and women that regularly exercise (Yonekura et al., 2007). Likewise, the dysregulation of tryptophan kynurenine metabolism has been associated with the development of metabolic syndrome (Oxenkrug, 2010), and in mice, this obesity vulnerability is gender-dependent (Uçeyler et al., 2010).

Tryptophan also exerts indirect effects on other neurotransmitters and several compounds from the central nervous system. L-tryptophan is converted into 5-hydroxy-tryptophan (5-HTP) by hydroxylases and then the 5-HTP is converted into 5-hydroxy-Tryptamin (5-HT) or serotonin by decarboxylases. Serotonin is converted into melatonin by the enzyme arylalkylamine N-acetyltransferase (AA-NAT), which constitutes the limiting step in melatonin synthesis (Vallone et al., 2007). This enzyme presents a marked circadian rhythm in all species that have been studied (Arendt, 1995), and is generally considered to be the most important factor for rhythmicity since it is rapidly induced via adrenergic stimulation, and rapidly inactivated, most likely by proteasomes (Klein, 2007). Decline of melatonin levels and MT1 receptor has been demonstrated on mammary tumor growth in aging rats (Hill et al., 2011).

In addition, the daily rhythmicity of melatonin secretion has been associated with sleep-wake cycle (Lien, 2003), but does not seem to be necessary for the regulation of the rat sleep-wake cycle under alternating light-dark photoperiod (Simon & Sugden, 2010). However alterations in dark-light cycle in rats demonstrate modifications of melatonin levels, food intake, feed efficiency, visceral adiposity, circadian rhythms along with increase in irritability and excitability (Wideman & Murphy, 2009). A review suggests the relation between night work and increase risk of chronic disease development (Wang et al., 2011). Tryptophan also had been associated with effects on other neurotransmitters. Oral dosing of tryptophan has been shown to increase dopamine, norepinephrine, and beta-endorphin levels (Richard et al., 2009). On the other hand, the dopamine and opioids systems are involved in the regulation of consumption of palatable foods (Vucetic et al., 2010).

Besides serotonin synthesis, tryptophan is also thought to modulate the turn-over of cortisol (Maes et al., 1989) as well as prolactin and growth hormone (Winokur et al., 1986), cognitive performance (Andolina et al., 2011), and memory (Lieben et al., 2004).

Acute intraperitoneal injection of a single dose of L-tryptophan (125mg/kg) decreases cortical spreading depression, a factor that can be associated with epilepsy and serotonin availability in the central nervous system (Trindade-Filho et al., 2009). An acute depletion of tryptophan in rats caused by a mixture of carbohydrate with tryptophan-free proteins can reduce the ratio TRP/LNAA within 2 and 4 h by 75 and 60%, respectively; without alterations of anxiety or depression behavior but with impairment of the capacity of object recognition (Lieben et al., 2004). There are studies showing that sensory system in children (Manjarrez et al., 2005) and animal (Manjarrez et al., 1994) with RGIU elevated plasma tryptophan levels and serotonin synthesis. Animals with growth retardation demonstrate altered expression of circadian rhythms and of food intake (Orozco-Solis et al., 2011). Given these data, tryptophan may be involved in metabolic disorders. Thus, tryptophan availability, serum and tissues melatonin and serotonin concentrations seem to have been associated with alterations in circadian activities and in metabolic activities that predispose to the emergence of chronic disease, food changes and obesity.

These metabolic pathways associated directly or indirectly with tryptophan availability have induced several lines of research on the depletion (Geeraerts et al., 2011; Steinberg et al., 1992; Lieben et al., 2004) or the supplementation (Fazzolari-Nesci et al., 1992; Sánchez et al., 2008; Kollman et al., 2008; Sarwar, 2001) acute or chronic with tryptophan.

3. Tryptophan in Breast Milk, an Antioxidant which Can Drive the Infants' Circadian Rhythm?

The daily requirements of nutrients for baby are based on their contents in human milk. Commercial infant formulas are generally designed to contain the major nutrients, but understanding of the amino acid requirements of baby from birth to weaning is incomplete (Chuang et al., 2005). Published amino acid requirements for the neonate are based on the assumption that human milk will meet all of the amino acid requirements of the infant. However, the nutritive value of a protein depends not only upon its content and balance of essential amino acids but also upon the availability of the amino acids to the organism growth (Heger & Frydrych, 1985). The deficit of protein or essential amino acids is associated with growth reduction during lactation and animals growth (Orozco-Solis et al., 2011). But animals submitted to a reduced intake in essential amino acids show decreased of oxidation ratio, greater efficiency of conservation of these amino acids and reduced catabolism (Young & Marchini, 1990). The major objective of this alteration of enzymatic kinetic is to increase the facility of driving amino acids for protein synthesis.The tryptophan is the amino acid most inefficiently used for protein synthesis, i.e., over 50% of absorbed tryptophan is apparently not used for protein synthesis. In this way, the utilization of tryptophan by different pathways may vary on physiological function and pathological status. Although the proportion of tryptophan utilized for protein synthesis varies within a great range, the average proportion of protein with tryptophan in the body is 1.2g/100g of protein which is much lower than other indispensable amino acids. The reasons for this discrepancy can be due to the physiological status, to the proportion of others amino acids, as well as to protein quantity and quality and growing period (Lin et al., 1986). In animals, the requirement and the proportion of amino acids for growth have been demonstrated in previous studies (Benevenga et al., 1994). In 1956, Gupta and Elvehjem (Heger & Frydrych, 1985) have shown that tryptophan level is important to the growth of weanling rats, as well as for food intake, i e, low levels is associated with low food intake. This same study demonstrates that tryptophan availability is different according to the origins of proteins whose availability of casein protein is 80,9%. But the availability of tryptophan can also change according to other nutrients present in diet. Diet high in carbohydrates favours the insulin secretion and induces transfer of LNAAs ratio into muscle tissue producing a relatively high plasma TRP/LNAAs and enhances transfer of TRP into the brain (Fernstrom & Wurtman, 1979). Newborns and infants may be at risk of insufficient bioavailability of tryptophan for optimal synthesis of brain serotonin. In infants, glucagon and epinephrine (Girard & Pegorier, 1998) limits the insulin-induced flow of LNAAs into muscle tissue. Combined with the lower muscle mass of infants (12% compared with 20% for adults) and their higher protein intake per unit of body mass per day (breastfed infants: 2 g/kg; formula-fed infants: 2–4 g/kg; adults: 0.8–1.2 g/kg), infants must rely more heavily on the dietary balance of TRP/LNAAs to maintain adequate brain tryptophan uptake (Lien, 2003), whose amino acid transport is crucial for serotonin and other metabolites synthesis involved in the post-natal brain development. In newborn, the serotonin system has been associated with the regulation of sleep-wake cycle as well as with brain development (Steinberg 1992). More recently abnormality in serotonergic system has been implicated for sudden infant death syndrome (Duncan et al., 2010;). Human milk contains a higher content of tryptophan than do both soy and cow milk-based infant formulas (Sarwar, 2001). Tryptophan concentration of Human Milk protein is about 2.2–2.4% compared with 1.3% in cow. The difference in tryptophan content of Human Milk and infant formula is mainly due to the alpha-lactalbumin portion of whey protein. Cow milk has a whey:casein ratio of approximately 20:80, which is distinctly different from the 60:40 whey:casein ratio in mature Human Milk (Rudloff & Kunz, 1997). Feeding infants with lactalbumin–rich formulas resulted in higher plasma tryptophan concentrations (Steinberg et al., 1992; Sarwar, 2001; Fazzolari-Nesci et al., 1992) and higher sleep latency than infants fed with standard formulas (Steinberg et al., 1992). Differences in circadian sleep patterns have been associated with the maturation and changes in brain function (Yogman & Zeisel, 1985). In humans, the sleep-wake cycle appears to be consolidated around the third to sixth months of life (Pasqualini et al., 2003) and the fully established activity/rest rhythm of three-month-old breast-fed infants (Horne et al., 2004) showed significantly longer than of the formula-fed. Babies fed with infant formulas containing less than 13g protein/L but supplemented with 3 concentrations of tryptophan show less latency to sleep than in the supplemented formula groups and similar to that in the breastfed group (Steinberg et al., 1992). This fact can be partly associated with a greater bioavailability of tryptophan in breast milk compared with formulas-fed. In rats subjected to low protein diet during pregnancy Kalyanasundaram and Ramana Murthy (1981) report that this activity can be prolonged around 35 days of life. Babies initially exhibit ultradian rhythms in their sleep-wake and food intake cycles, waking frequently and feeding often (Sumovà et al., 2006). Differences in time spent in sleep and waking states are thought to influence the development of the central nervous system during infancy (Steinberg et al., 1992). In rodents, the mother feeds her pups and thus keeps them mostly active during the day. But adult rats, however, are active and consume food at night. During the weaning period, between P14 and P28, the pup's feeding and activity regimes apparently change. The shifts, though smaller, continue, together with a drastic change of the rhythm amplitude between P20 and P30, when the matured circadian system seems to have been established (Sumovà et al., 2006).

However, these processes need more research to understand specific mechanisms and effects of supplementation by tryptophan. Nascimento et al., (2013) have tried to use tryptophan supplementation to prevent the onset of metabolic syndrome subsequently to severe protein restriction. Low protein as well as control rat pups receiving a bolus of tryptophan up to weaning have shown alteration in circadian rhythms or general physiology in infancy and adulthood. This work underlines the need to foster understanding of the onset of circadian rhythms and of behavior in the rat pups to be able to assay the effect of tryptophan on rat physiology.

It has been shown that alpha-lactalbumin can increase plasma TRP/LNAA ratio up to 48% as compared to casein (Olivier et al., 2002). The synthesis of alpha-lactalbumin for mammary gland is essential for lactose and milk production. Alpha-lactalbumin is a protein essential in milk of all mammalian species (Minet-Ringuet et al., 2004) and determinant in lactose formation in transgenic animals (Lien, 2003). Lactose content provides the driving osmotic force behind the fluid formation of milk. On the other hand, the food restriction significantly reduces the synthesis and the secretion of casein and alfa-lactalbumin in rat cells. Acini cells from rats receiving dietary treatments with 10% protein diet have shown significantly lower rates of synthesis and secretion of alfa-lactalbumin (Sampson et al., 1986).The synthesis of proteins in the mammary gland of rat has been shown to be dependent on prolactin and placental lactogen and others lactogenic hormones (Rhoads et al., 2007). The lower levels of prolactin in mammary gland seem to compromise milk production and the growth of pups (Moretto et al., 2010) as well as low levels of amino acids (Rhoads et al., 2007). The secretion of prolactin is stimulated by serotonergic neurons terminating in the hypothalamus, and prolactin secretion is shown to be increased by L-tryptophan or serotonin agonists (Van De Kar et al., 1996). The concentration of tryptophan is variable along 24 hours in the milk of lactating mother. Maxima of tryptophan concentration are in the range of 80-90 μM/L and minima in the range of 45-55 μM/L (Cubero et al., 2005). These values are variable between samples and lactating stages (Chuang et al., 2005) and also varies in function of nutritional status of mothers during pregnancy and after delivery (Lonnerdal, 1986), but the major concentrations of amino acids are in colostrum milk (Chuang et al., 2005). Hypoprotein diet (6%) in rats during lactation reduces food intake and impaired prolactin levels, growth gland mammary tissue (Moretto et al., 2010), consequently milk flux and the growth of pups. Previous studies have demonstrated low plasma tryptophan concentration in premature infant and in infants with protein-caloric malnutrition (Ghisolfi et al., 1978). Newborn infants are at greater risk of oxidative stress because of an imbalance between high oxidant loads and immature antioxidant defenses (Lavoie & Chessex, 1997). In human milk, L-tryptophan has been recently shown to display anti-oxidant properties (Tsopmo et al., 2009) whose advantage seem higher that E and C vitamins. In addition, newborns with RGIU have greater exposure to oxidants reactive species and antioxidants components shown beneficial effects to development (Gitto et al., 2009) Similarly rats submitted to the low-protein diet or restriction diet exhibit high levels of stress hormone. Rats fed low protein diet (8-9%) with more (0.1%) or less tryptophan (0.058% TRYP) levels in diet display different susceptibility to convulsions, a conditioning associated with the lower rate of brain serotonin synthesis, important for development and maturation of GABAergic inhibitory cortical interneurons (Feria-Velasco et al., 2008). Weanling rats, receiving for 2 weeks soy-based infant formulas supplemented by tryptophan, show reduced food consummation and low body weight gain, elevation of tryptophan in plasma and brain and tryptophan 5-hydroxyindoles in brain with an increase proportional to tryptophan concentration (Sarwar, 2001). However, another study with test formulas containing graded levels of L-tryptophan (0.1 or 0.2%) does not report any effect on rat growth (according to the protein efficiency ratio, Net Protein ratio and relative Net Protein ratio), but has found increases ($p < 0.05$) in the concentrations of tryptophan in the plasma and brain, of serotonin and 5-HIAA in brain of rats (Sawar, 2001). Recently it has been demonstrated that weaning time influence the circadian rhythms of feeding in the long term (dos Santos Oliveira et al., 2011). The behavioral satiety sequence in adulthood indicates that the animals weaned later were quicker to reach satiety, consume less food in less time for eating (dos Santos Oliveira et al., 2011) than early weaned rats. Among other factors, offspring with a shorter suckling period can reduce tryptophan availability and consequently decrease formation of 5-hydroxytryptamine in brain areas implicated in food consummation modulation. Tryptophan requirements for animals are defined as the amount of tryptophan required to maximize growth rate. Thus, a state of deficiency begins when dietary tryptophan content does not permit the highest growth rate. This fact is also common in others essential amino acids. Ettle and Roth showed that piglets are able to detect tryptophan deficiency and develop an aversion against a tryptophan-deficient diet. According to Heger & Frydrych (1985), tryptophan amount for the growing and maintenance for rats is of 1,5g/kg diet and of 8mg/kg/day, respectively. In pigs, the effect of tryptophan deficiency on growth is mainly caused by a reduction of feed intake (Eder et al., 2003). Tryptophan and cysteine concentrations in cow milk are approximately one half of those found in human milk, when expressed as a portion of total protein (Heine et al., 1991). Two milk formulas are used in Spain for newborns with a day/wake formula containing higher proteins, cytidine, guanosine and inosine whilst the sleep promoting formula contained more L-tryptophan (Rivero et al., 2005). These formulas have been tested on rat pups (Puigjaner et al., 2007) but not for the expression of clock genes in different tissues and not on long-term health consequences. In addition to our knowledge the formula is not public and the milk is not used in neonatal intensive care units. To explore nutritiornal questions regarding the deficiency of protein and/or amino acids supplementation in a state of malnutrition (Lopes De Souza et al., 2008; Orozco-Solis et al., 2011; Delamaire et al., 2008) models of rats fed with low-protein diet have been designed. Preterm infants born with a very low birth weight suffer frequently of perinatal malnutrition, factor involved for developing obesity and metabolic disorders in the adulthood. The effects of malnutrition on short-term and long-term growth of rat pups are well-established since the seminal work of Mccance (1962) and the importance of circadian rhythms in gut physiology are current trends on rat and human physiology (Pardini et al., 2005; Sanger et al., 2008; Bertrand & Bertrand, 2010). Malnutrition is believed to imprint on the long-term the mammalian physiology when experienced shortly after birth (between birth and weaning, in rat from day-1 to day-20). The postnatal period in rats is critic for development of tryptophan hydroxylase (TPH) and serotonin transporter (SERT) in multiple brainstem nuclei from P2 to P21 (Liu et al., 2011). Thus this period is a time window during which appropriate nutritional corrections can be applied with success in order to revert or smooth the deleterious effects of malnutrition over the short or long term. We are exploring the detrimental effects of early malnutrition on the circadian system (Orozco-Solis et al., 2011) after reporting serotonin dysregulation in a model of under nourished rats (Lopes De Souza et al., 2008). Dysregulation of serotonin pathways during restricted nutrition in rat pups has been demonstrated on the desensitization of 5-HT1B receptors (Lopes De Souza et al., 2008) and offspring hyperphagic to 35th old age (Orozco Solis et al 2010). In parallel, we have developed non-invasive technology to explore gut functions in preterm neonates (Kaeffer et al., 2007a) allowing *in vitro* double-checking on gastric and intestinal cells of babies or rat pups of short-term effects of a supplementation by tryptophan.The circadian expression pattern of several genes regulating food intake is also altered in undernourished rat pups. The relationship with clock's functioning has been explored on period1 and clock genes (Orozco-Solis et al., 2011). The existence of clock genes in the digestive tract of adult humans (Pardini et al., 2005) and preterm infants (Kaeffer et al., 2007b ; Kaeffer et al., 2011) has been demonstrated. The inducing and synchronizing capacity of feeding on circadian peripheral clocks remain little explored in neonates (Kennaway & Wright, 2002; Kaeffer et al., 2007b; Kaeffer et al., 2011) but it has been the subject of many studies in rat, with the recent demonstration of an entraining effect of maternal breast feeding on the circadian rhythmicity of the liver's clock of small rats (Sladek et al., 2006) as well as the alignment between circadian regulation of intestine relatively to the central clock located in the suprachiasmatic nucleus. Recently, we have manipulated the long-term sequelae of early-life programming by undernutrition and found that cultured cells retain circadian clock dysregulation (Nascimento et al., 2013).

Conclusion and Perspectives

As tryptophan is an essential amino acid, a diet deficient in tryptophan, may be implicated in the development of the central nervous system and the acquisition of circadian rhythms. The disruption of rhythms can induce a number of behavioural abnormalities, probably mediated through the decreased formation of 5-hydroxytryptamine or other metabolites of tryptophan. Future studies investigating the role of an excess or of a deficit of tryptophan and the consequences on infant development and metabolism are deeply needed.

References

Andolina D, Conversi D, Cabib S, Trabalza A, Ventura R, Puglisi-Allegra S, Pascucci T. 5-Hydroxytryptophan during critical postnatal period improves cognitive performances and promotes dendritic spine maturation in genetic mouse model of phenylketonuria. Int J Neuropsychopharmacol. 2011 May;14(4):479-89. doi:10.1017/S1461145710001288. Epub 2010 Nov 1. PubMed PMID: 21040618; PubMed Central PMCID: PMC 3110346.

Arendt J. Melatonin and human rhythms. Chronobiol Int. 2006;23(1-2):21-37. Review. PubMed PMID: 16687277.

Arendt J. Melatonin and the mammalian pineal gland. Springer Science & Business Media, 1995.

Baker DH. Comparative nutrition and metabolism: explication of open questions with emphasis on protein and amino acids. Proc Natl Acad Sci U S A. 2005 Dec 13;102(50):17897-902. Epub 2005 Dec 2. Review. PubMed PMID: 16326801; PubMedCentral PMCID: PMC1312430.

Benevenga NJ, Gahl MJ, Crenshaw TD, Finke MD. Protein and amino acid requirements for maintenance and amino acid requirements for growth of laboratory rats. J Nutr. 1994 Mar;124(3):451-3. PubMed PMID: 8120665.

Bertazzo A, Comai S, Brunato I, Zancato M, Costa CVL. The content of protein and non-protein (free and protein-bound) tryptophan in Theobroma cacao beans. *Food Chemistry,* 124 (2011) 93–96

Bertrand PP, Bertrand RL. Serotonin release and uptake in the gastrointestinal tract. Auton Neurosci. 2010 Feb 16;153(1-2):47-57. doi: 10.1016/j.autneu. 2009.08.002. Epub 2009 Sep 2. Review. PubMed PMID: 19729349.

Borjigin J, Li X, Snyder SH. The pineal gland and melatonin: molecular and pharmacologic regulation. Annu Rev Pharmacol Toxicol. 1999;39:53-65. Review. PubMed PMID: 10331076.

Bray MS, Young ME. The role of cell-specific circadian clocks in metabolism and disease. Obes Rev. 2009 Nov;10 Suppl 2:6-13. doi:10.1111/j.1467-789X.2009.00684.x. Review. PubMed PMID: 19849797.

Chan JS, Snoeren EM, Cuppen E, Waldinger MD, Olivier B, Oosting RS. The serotonin transporter plays an important role in male sexual behavior: a study in serotonin transporter knockout rats. J Sex Med. 2011 Jan;8(1):97-108. doi:10.1111/j.1743-6109.2010.01961.x. PubMed PMID: 20704641.

Chessex P, Watson C, Kaczala GW, Rouleau T, Lavoie ME, Friel J, Lavoie JC. Determinants of oxidant stress in extremely low birth weight premature infants. Free Radic Biol Med. 2010 Nov 15;49(9):1380-6. doi:10.1016/j.freeradbiomed.2010.07.018. Epub 2010 Aug 5. PubMed PMID: 20692335.

Chowdhury I, Sengupta A, Maitra SK. Melatonin: fifty years of scientific journey from the discovery in bovine pineal gland to delineation of functions in human. Indian J Biochem Biophys. 2008 Oct;45(5):289-304. Review. PubMed PMID:19069840.

Chuang CK, Lin SP, Lee HC, Wang TJ, Shih YS, Huang FY, Yeung CY. Free amino acids in full-term and pre-term human milk and infant formula. J Pediatr Gastroenterol Nutr. 2005 Apr;40(4):496-500. PubMed PMID: 15795601.

Coates MD, Mahoney CR, Linden DR, Sampson JE, Chen J, Blaszyk H, Crowell MD, Sharkey KA, Gershon MD, Mawe GM, Moses PL. Molecular defects in mucosal serotonin content and decreased serotonin reuptake transporter in ulcerative colitis and irritable bowel syndrome. Gastroenterology. 2004 Jun;126(7):1657-64. PubMed PMID:15188158.

Colmenares JL, Wurtman RJ, Fernstrom JD. Effects of ingestion of a carbohydrate-fat meal on the levels and synthesis of 5-hydroxyindoles in various regions of the rat central nervous system. *J. Neurochem.,* 1975 Dec;25(6):825-9. PubMed PMID: 1082010.

Côté F, Fligny C, Bayard E, Launay JM, Gershon MD, Mallet J, Vodjdani G. Maternal serotonin is crucial for murine embryonic development. Proc Natl Acad Sci U S A. 2007 Jan 2;104(1):329-34. Epub 2006 Dec 20. PubMed PMID: 17182745; PubMed Central PMCID: PMC1713169.

Côté F, Fligny C, Fromes Y, Mallet J, Vodjdani G. Recent advances in understanding serotonin regulation of cardiovascular function. Trends Mol Med. 2004 May;10(5):232-8. Review. PubMed PMID: 15121050.

Cubero, J. et al. Ion-exchange HPLC assay of tryptophan in breast milk of mothers with three months breast-feeding. *Biogenic amines,* v. 19, n. 3, p. 171-175, 2005.

Deiró, Teresa Cristina Bomfim de Jesus et al. Neonatal exposure to citalopram, a serotonin selective reuptake inhibitor, programs a delay in the reflex ontogeny in rats. Arquivos de neuro-psiquiatria, v. 66, n. 3B, p. 736-740, 2008.

Delamaire E Hoebler C, Michel C, Kaeffer B et al., SFRP-02- Recherche expérimentale – Effect mémoire d□une alimentation hyperprotéique chez le raton. *Archives Pédiatrie,* 2008; 15:1031-1038.

Dileo HA.; Reiter RJ.; Tallaferro, DH. Chronobiology, melatonin, and sleep in infants and children. *Pediatric nursing,* v. 28, n. 1, p. 35-39, 2002.

dos Santos L, de Lima DP, da Silva AA, da Silva MC, de Souza SL, Manhães-de-Castro R. Early weaning programs rats to have a dietary preference for fat and palatable foods in adulthood. Behav Processes. 2011 Jan;86(1):75-80. doi:10.1016/j.beproc.2010.09.005. Epub 2010 Oct 8. PubMed PMID: 20934490.

Duncan JR, Paterson DS, Hoffman JM, Mokler DJ, Borenstein NS, Belliveau RA, Krous HF, Haas EA, Stanley C, Nattie EE, Trachtenberg FL, Kinney HC. Brainstem serotonergic deficiency in sudden infant death syndrome. JAMA. 2010 Feb3;303(5):430-7. doi: 10.1001/jama.2010.45. PubMed PMID: 20124538; PubMed Central PMCID: PMC3242415.

Eder K, Nonn H, Kluge H, Peganova S. Tryptophan requirement of growing pigs at various body weights. J Anim Physiol Anim Nutr (Berl). 2003 Oct;87(9-10):336-46. PubMed PMID: 14507416.

Ednick M, Cohen AP, McPhail GL, Beebe D, Simakajornboon N, Amin RS. A review of the effects of sleep during the first year of life on cognitive, psychomotor, and temperament development. Sleep. 2009 Nov;32(11):1449-58. Review. PubMed PMID:19928384; PubMed Central PMCID: PMC2768951.

Ellinger, A; Flamand, C. Über synthetisch gewonnenes Tryptophan und einige seiner Derivate. Hoppe-Seyler□s Zeitschrift für physiologische Chemie, v. 55, n. 1, p. 8-24, 1908..

Esteban S, Nicolaus C, Garmundi A, Rial RV, Rodríguez AB, Ortega E, Ibars CB. Effect of orally administered L-tryptophan on serotonin, melatonin, and the innate immune response in the rat. Mol Cell Biochem. 2004 Dec;267(1-2):39-46. PubMed PMID: 15663184.

Fazzolari-Nesci A, Domianello D, Sotera V, Räihä NC. Tryptophan fortification of adapted formula increases plasma tryptophan concentrations to levels not different from those found in breast-fed infants. *J. Pediatr. Gastroenterol. Nutr.,* 1992 May;14(4):456-9. PubMed PMID: 1517950.

Feria-Velasco A, Mena-Munguía S, Cárabez-Torres J, Gómez-Medrano A, Recéndiz-Hurtado F, Orozco-Suárez S, Beas-Zárate C. Low tryptophan and protein in the diet during development increase the susceptibility to convulsions in adult rats. *Neurochem. Res.,* 2008 Aug;33(8):1484-91. doi: 10.1007/s11064-008-9613-0. Epub 2008 Feb 14. PubMed PMID: 18273702.

Fernstrom JD, Wurtman RJ, Hammarstrom-Wiklund B, Rand WM, Munro HN, Davidson CS. Diurnal variations in plasma concentrations of tryptophan, tryosine, and other neutral amino acids: effect of dietary protein intake. *Am. J. Clin. Nutr.,* 1979 Sep;32(9):1912-22. PubMed PMID: 573061.

Fisher SP, Sugden D. Endogenous melatonin is not obligatory for the regulation of the rat sleep-wake cycle. Sleep. 2010 Jun;33(6):833-40. PubMed PMID: 20550025; PubMed Central PMCID: PMC2881717.

Folch J, Lees M, Sloane Stanley GH. A simple method for the isolation and purification of total lipides from animal tissues. *J. Biol. Chem.,* 1957 May;226(1):497-509. PubMed PMID: 13428781.

Fonken LK, Workman JL, Walton JC, Weil ZM, Morris JS, Haim A, Nelson RJ. Light at night increases body mass by shifting the time of food intake. Proc Natl Acad Sci U S A. 2010 Oct 26;107(43):18664-9. doi: 10.1073/pnas.1008734107. Epub 2010 Oct 11. PubMed PMID: 20937863; PubMed Central PMCID:PMC2972983.

Fuller PM, Lu J, Saper CB. Differential rescue of light- and food-entrainable circadian rhythms. Science. 2008 May 23;320(5879):1074-7. doi:10.1126/science.1153277. PubMed PMID: 18497298; PubMed Central PMCID: PMC3489954.

Geeraerts B, Van Oudenhove L, Boesmans W, Vos R, Vanden Berghe P, Tack J. Influence of acute tryptophan depletion on gastric sensorimotor function in humans. *Am. J. Physiol. Gastrointest Liver Physiol.,* 2011 Feb;300(2):G228-35. Doi: 10.1152/ajpgi.00020.2010. Epub 2010 Sep 30. PubMed PMID: 20884888.

Gershon MD, Tack J. The serotonin signaling system: from basic understanding to drug development for functional GI disorders. Gastroenterology. 2007 Jan;132(1):397-414. Review. PubMed PMID: 17241888.

Ghisolfi J, Charlet P, Ser N, Salvayre R, Thouvenot JP, Duole C. Plasma free amino acids in normal children and in patients with proteinocaloric malnutrition: fasting and infection. *Pediatr. Res.,* 1978 Sep;12(9):912-7. PubMed PMID: 101961.

Gill RK, Pant N, Saksena S, Singla A, Nazir TM, Vohwinkel L, Turner JR, Goldstein J, Alrefai WA, Dudeja PK. Function, expression, and characterization of the serotonin transporter in the native human intestine. Am J Physiol Gastrointest Liver Physiol. 2008 Jan;294(1):G254-62. Epub 2007 Nov 8. PubMed PMID: 17991706.

Girard J, Pégorier JP. An overview of early post-partum nutrition and metabolism. *Biochem. Soc. Trans.,* 1998 May;26(2):69-74. Review. PubMed PMID: 9649722.

Gitto E, Pellegrino S, Gitto P, Barberi I, Reiter RJ. Oxidative stress of the newborn in the pre- and postnatal period and the clinical utility of melatonin. J Pineal Res. 2009 Mar;46(2):128-39. doi: 10.1111/j.1600-079X.2008.00649.x. Epub 2008 Nov 19. Review. PubMed PMID: 19054296.

Green CB, Menaker M. Circadian rhythms. Clocks on the brain. Science. 2003 Jul 18;301(5631):319-20. Epub 2003 Jul 3. PubMed PMID: 12843400.

Grigor MR, Allan JE, Carrington JM, Carne A, Geursen A, Young D, Thompson MP, Haynes EB, Coleman RA. Effect of dietary protein and food restriction on milk production and composition, maternal tissues and enzymes in lactating rats. *J. Nutr.,* 1987 Jul;117(7):1247-58. PubMed PMID: 3612304.

Heger J, Frydrych Z. Efficiency of utilization of essential amino acids in growing rats at different levels of intake. Br J Nutr. 1985 Sep;54(2):499-508. PubMed PMID: 4063333.

Heine WE, Klein PD, Reeds PJ. The importance of alpha-lactalbumin in infant nutrition. *J. Nutr.,* 1991 Mar;121(3):277-83. Review. PubMed PMID: 2002399.

Hill SM, Cheng C, Yuan L, Mao L, Jockers R, Dauchy B, Frasch T, Blask DE. Declining melatonin levels and MT1 receptor expression in aging rats is associated with enhanced mammary tumor growth and decreased sensitivity to melatonin. Breast Cancer Res Treat. 2011 May;127(1):91-8. Doi: 10.1007/s10549-010-0958-0. Epub 2010 Jun 12. PubMed PMID: 20549340.

Hoogerwerf WA. Role of biological rhythms in gastrointestinal health and disease. *Rev. Endocr. Metab. Disord.,* 2009 Dec;10(4):293-300. doi:10.1007/s11154-009-9119-3. Review. PubMed PMID: 19798581.

Hopkins FG, Cole SW. A contribution to the chemistry of proteids: Part I. A preliminary study of a hitherto undescribed product of tryptic digestion. *J. Physiol.,* 1901 Dec 23;27(4-5):418-28. PubMed PMID: 16992614; PubMed Central PMCID:PMC1540554.

Horne RS, Parslow PM, Ferens D, Watts AM, Adamson TM. Comparison of evoked arousability in breast and formula fed infants. *Arch. Dis. Child.,* 2004 Jan;89(1):22-5. PubMed PMID: 14709496; PubMed Central PMCID: PMC1755888.

Horne RS, Parslow PM, Ferens, D, watts AM, Adamson TM. Comparison of evoked arousability in breast and formula fed infants. *Arch. Dis. Child.,* 2004; 89:22-25.

Huang Z, Liu T, Borjigin J. N-terminal residues regulate proteasomal degradation of AANAT. *J. Pineal. Res.,* 2010 Apr;48(3):290-6. doi:10.1111/j.1600-079X.2010.00753.x. Epub 2010 Mar 4. PubMed PMID: 20210853.

Jin X, Shearman LP, Weaver DR, Zylka MJ, de Vries GJ, Reppert SM. A molecular mechanism regulating rhythmic output from the suprachiasmatic circadian clock. *Cell,* 1999 Jan 8;96(1):57-68. PubMed PMID: 9989497.

Jones RS. Tryptamine: a neuromodulator or neurotransmitter in mammalian brain? *Prog. Neurobiol.,* 1982;19(1-2):117-39. PubMed PMID: 6131482.

Kaeffer B, des Robert C, Alexandre-Gouabau MC, Pagniez A, Legrand A, Amarger V, Küster A, Piloquet H, Champ M, le Huërou-Luron I, Rozé JC. Recovery of exfoliated cells from the gastrointestinal tract of premature infants: a new tool to perform "noninvasive biopsies?". *Pediatr. Res.,* 2007(a) Nov;62(5):564-9. PubMed PMID: 17805197.

Kaeffer B, Legrand A, Moyon T, Frondas-Chauty A, Billard H, Guzman-Quevedo O, Darmaun D, Rozé JC. Non-invasive exploration of neonatal gastric epithelium by using exfoliated epithelial cells. PLoS One.

2011;6(10):e25562. Doi: 10.1371/journal.pone.0025562. Epub 2011 Oct 18. PubMed PMID: 22028779; PubMed Central PMCID: PMC3196493.

Kaeffer B. Acquisition of Circadian Rhythms by Neonates: A Review on the Potentials of Exfoliated Cells to Study Gastrointestinal Tract Maturation and Clockwork. Chapter 5, pp. 99-116. In: Circadian Rhythms and Health Research Trends ISBN 978-1-60021-875-x Editor: Alfredo V. Lignelli, pp 206, Novapublishers; 2007(b)

Kalyanasundaram S, Ramanamurthy PS. Effect of undernutrition on tryptophan and tyrosine hydroxylases in the developing rat brain. *J. Neurochem.,* 1981 Apr;36(4):1580-2. PubMed PMID: 6114988.

Kanai Y, Endou H. Functional properties of multispecific amino acid transporters and their implications to transporter-mediated toxicity. *J. Toxicol. Sci.,* 2003 Feb;28(1):1-17. Review. PubMed PMID: 12696180.

Kaper T, Looger LL, Takanaga H, Platten M, Steinman L, Frommer WB. Nanosensor detection of an immunoregulatory tryptophan influx/kynurenine efflux cycle. *PLoS Biol.,* 2007 Oct;5(10):e257. PubMed PMID: 17896864; PubMed Central PMCID: PMC1988858.

Kennaway DJ, Wright H. Melatonin and circadian rhythms. Curr Top Med Chem. 2002 Feb;2(2):199-209. Review. PubMed PMID: 11899101.

Khaliq S, Haider S, Ahmed SP, Perveen T, Haleem DJ. Relationship of brain tryptophan and serotonin in improving cognitive performance in rats. *Pak. J. Pharm. Sci.,* 2006 Jan;19(1):11-5. PubMed PMID: 16632446.

Khaliq S, Haider S, Ahmed SP, Perveen T, Haleem DJ. Relationship of brain tryptophan and serotonin in improving cognitive performance in rats. Pak *J. Pharm. Sci.,* 2006 Jan;19(1):11-5. PubMed PMID: 16632446.

Klein DC, Coon SL, Roseboom PH, Weller JL, Bernard M, Gastel JA, Zatz M, Iuvone PM, Rodriguez IR, Bégay V, Falcón J, Cahill GM, Cassone VM, Baler R. The melatonin rhythm-generating enzyme: molecular regulation of serotonin N-acetyltransferase in the pineal gland. Recent Prog Horm Res. 1997;52:307-57; discussion 357-8. Review. PubMed PMID: 9238858.

Klein DC. Arylalkylamine N-acetyltransferase: "the Timezyme". J Biol Chem. 2007 Feb 16;282(7):4233-7. Epub 2006 Dec 12. Review. PubMed PMID: 17164235.

Kollmann MT, Locher M, Hirche F, Eder K, Meyer HH, Bruckmaier RM. Effects of tryptophan supplementation on plasma tryptophan and related hormone levels in heifers and dairy cows. Domest Anim Endocrinol. 2008 Jan;34(1):14-24. Epub 2006 Oct 19. PubMed PMID: 17074461.

Krueger JM, Rector DM, Roy S, Van Dongen HP, Belenky G, Panksepp J. Sleep as a fundamental property of neuronal assemblies. Nat Rev Neurosci. 2008 Dec;9(12):910-9. doi: 10.1038/nrn2521. Epub 2008 Nov 5. Review. PubMed PMID: 18985047; PubMed Central PMCID: PMC2586424.

Lavoie JC, Chessex P. Gender and maturation affect glutathione status in human neonatal tissues. *Free Radic. Biol. Med.,* 1997;23(4):648-57. PubMed PMID: 9215810.

Leibowitz SF, Alexander JT. Hypothalamic serotonin in control of eating behavior, meal size, and body weight. *Biol Psychiatry,* 1998 Nov 1;44(9):851-64. Review. PubMed PMID: 9807640.

Li S, Lin JD. Molecular control of circadian metabolic rhythms. J Appl Physiol (1985).2009Dec;107(6):1959-64. Doi: 10.1152/japplphysiol. 00467.2009. Epub 2009 Jul 2. Review. PubMed PMID: 19574505.

Lieben CK, van Oorsouw K, Deutz NE, Blokland A. Acute tryptophan depletion induced by a gelatin-based mixture impairs object memory but not affective behavior and spatial learning in the rat. *Behav. Brain Res.,* 2004 May 5;151(1-2):53-64. PubMed PMID: 15084421.

Lien EL. Infant formulas with increased concentrations of alpha-lactalbumin. Am J Clin Nutr. 2003 Jun;77(6):1555S-1558S. Review. PubMed PMID: 12812154.

Lin FD, Smith TK, Bayley HS. A role for tryptophan in regulation of protein synthesis in porcine muscle. *J. Nutr.,* 1988 Apr;118(4):445-9. PubMed PMID: 3357059.

Liu Q, Wong-Riley MT. Postnatal changes in tryptophan hydroxylase and serotonin transporter immunoreactivity in multiple brainstem nuclei of the rat: implications for a sensitive period. *J. Comp Neurol.,* 2010 Apr 1;518(7):1082-97. doi: 10.1002/cne.22265. PubMed PMID: 20127812; PubMed Central PMCID: PMC3694345.

Liu T, Borjigin J. Relationship between nocturnal serotonin surge and melatonin onset in rodent pineal gland. J Circadian Rhythms. 2006 Sep 27;4:12.PubMed PMID: 17005040; PubMed Central PMCID: PMC 1592123.

Lönnerdal B. Effects of maternal dietary intake on human milk composition. *J. Nutr.,* 1986 Apr;116(4):499-513. Review. PubMed PMID: 3514820.

Lopes de Souza S, Orozco-Solis R, Grit I, Manhães de Castro R, Bolaños-Jiménez F. Perinatal protein restriction reduces the inhibitory action of serotonin on food intake. *Eur. J. Neurosci.,* 2008 Mar;27(6):1400-8.

doi:10.1111/j.1460-9568.2008.06105.x. Epub 2008 Mar 7. PubMed PMID: 18331342.

Maes M, Vandevelde R, Suy E. Influences on cortisol and noradrenergic turnover of healthy controls and depressed patients during L-tryptophan loading. *J. Affect Disord.,* 1989 Sep-Oct;17(2):173-82. PubMed PMID: 2549110.

Malek ZS, Dardente H, Pevet P, Raison S. Tissue-specific expression of tryptophan hydroxylase mRNAs in the rat midbrain: anatomical evidence and daily profiles. *Eur. J. Neurosci.,* 2005 Aug;22(4):895-901. PubMed PMID: 16115212.

Manhães de Castro R, Barreto Medeiros JM, Mendes da Silva C, Ferreira LM, Guedes RC, Cabral Filho JE, Costa JA. Reduction of intraspecific aggression in adult rats by neonatal treatment with a selective serotonin reuptake inhibitor. *Braz J. Med. Biol. Res.,* 2001 Jan;34(1):121-4. PubMed PMID: 11151037.

Manjarrez G, Cisneros I, Herrera R, Vazquez F, Robles A, Hernandez J. Prenatal impairment of brain serotonergic transmission in infants. *J. Pediatr.,* 2005 Nov;147(5):592-6. PubMed PMID: 16291347.

Manjarrez GG, Chagoya GG, Hernández J. Early nutritional changes modify the kinetics and phosphorylation capacity of tryptophan-5-hydroxylase. Int *J. Dev. Neurosci.,* 1994 Dec;12(8):695-702. PubMed PMID: 7747596.

Markus CR, Olivier B, de Haan EH. Whey protein rich in alpha-lactalbumin increases the ratio of plasma tryptophan to the sum of the other large neutral amino acids and improves cognitive performance in stress-vulnerable subjects. *Am. J. Clin. Nutr.,* 2002 Jun;75(6):1051-6. PubMed PMID: 12036812.

MCCANCE RA. Food, growth, and time. Lancet. 1962 Oct 6;2(7258):671-6. PubMed PMID: 13932032.

Mendoza J, Pévet P, Challet E. High-fat feeding alters the clock synchronization to light. *J. Physiol.,* 2008 Dec 15;586(Pt 24):5901-10. doi:10.1113/jphysiol.2008.159566. Epub 2008 Oct 20. PubMed PMID: 18936083; PubMed Central PMCID: PMC2655413.

Meyer T, Brinck U. Differential distribution of serotonin and tryptophan hydroxylase in the human gastrointestinal tract. Digestion. 1999 Jan-Feb;60(1):63-8. PubMed PMID: 9892800.

Meyer-Bernstein EL, Morin LP. Differential serotonergic innervation of the suprachiasmatic nucleus and the intergeniculate leaflet and its role in circadian rhythm modulation. *J. Neurosci.,* 1996 Mar 15;16(6):2097-111. PubMed PMID: 8604054.

Minet-Ringuet J, Le Ruyet PM, Tomé D, Even PC. A tryptophan-rich protein diet efficiently restores sleep after food deprivation in the rat. *Behav. Brain Res.,* 2004 Jul 9;152(2):335-40. PubMed PMID: 15196801.

Moretto VL, Ballen MO, Gonçalves TS, Kawashita NH, Stoppiglia LF, Veloso RV, Latorraca MQ, Martins MS, Gomes-da-Silva MH. Low-Protein Diet during Lactation and Maternal Metabolism in Rats. ISRN Obstet Gynecol. 2011;2011:876502. doi: 10.5402/2011/876502. Epub 2010 Oct 20. PubMed PMID: 21637364; PubMed Central PMCID: PMC3101884.

Mukai M, Tischkau SA. Effects of tryptophan photoproducts in the circadian timing system: searching for a physiological role for aryl hydrocarbon receptor. *Toxicol. Sci.,* 2007 Jan;95(1):172-81. Epub 2006 Oct 4. PubMed PMID: 17020875.

Nakahata Y, Grimaldi B, Sahar S, Hirayama J, Sassone-Corsi P. Signaling to the circadian clock: plasticity by chromatin remodeling. *Curr. Opin. Cell Biol.,* 2007 Apr;19(2):230-7. Epub 2007 Feb 20. Review. PubMed PMID: 17317138.

Nascimento E, Guzman-Quevedo O, Delacourt N, da Silva Aragão R, Perez-Garcia G, de Souza SL, Manhães-de-Castro R, Bolaños-Jiménez F, Kaeffer B. Long-lasting effect of perinatal exposure to L-tryptophan on circadian clock of primary cell lines established from male offspring born from mothers fed on dietary protein restriction. PLoS One. 2013;8(2):e56231. doi: 10.1371/journal.pone.0056231. Epub 2013 Feb 27. PubMed PMID: 23460795; PubMed Central PMCID: PMC3584092.

Olivier B, Broersen LM, Slangen JL. 5-HT3 receptor ligands lack discriminative stimulus properties. *Prog. Neuropsychopharmacol. Biol. Psychiatry,* 2002Apr;26(3):463-71. PubMed PMID: 11999896.

Orozco-Solís R, Matos RJ, Lopes de Souza S, Grit I, Kaeffer B, Manhães de Castro R, Bolaños-Jiménez F. Perinatal nutrient restriction induces long-lasting alterations in the circadian expression pattern of genes regulating food intake and energy metabolism. *Int. J. Obes.,* (Lond). 2011 Jul;35(7):990-1000. doi:10.1038/ijo.2010.223. Epub 2010 Nov 9. PubMed PMID: 21060311.

Oxenkrug GF. Metabolic syndrome, age-associated neuroendocrine disorders, and dysregulation of tryptophan-kynurenine metabolism. *Ann. N Y Acad. Sci.,* 2010 Jun;1199:1-14. doi: 10.1111/j.1749-6632.2009.05356.x. Review. PubMed PMID: 20633104.

Panda S, Antoch MP, Miller BH, Su AI, Schook AB, Straume M, Schultz PG, Kay SA, Takahashi JS, Hogenesch JB. Coordinated transcription of key pathways in the mouse by the circadian clock. *Cell,* 2002 May 3;109(3):307-20. PubMed PMID:12015981.

Pardini L, Kaeffer B, Trubuil A, Bourreille A, Galmiche JP. Human intestinal circadian clock: expression of clock genes in colonocytes lining the crypt. Chronobiol Int. 2005;22(6):951-61. PubMed PMID: 16393700.

Pasqualini, T. Implicancias de la restricción de crecimiento intrauterino (RCIU) sobre las enfermedades metabólicas. Mol Cancer Ther, v. 2, p. 265-71, 2003.

Penedo LA, Oliveira-Silva P, Gonzalez EM, Maciel R, Jurgilas PB, Melibeu Ada C, Campello-Costa P, Serfaty CA. Nutritional tryptophan restriction impairs plasticity of retinotectal axons during the critical period. Exp Neurol. 2009 May;217(1):108-15. doi: 10.1016/j.expneurol.2009.01.021. Epub 2009 Feb 10. PubMed PMID: 19416666.

Penkova NI, Baltadjiev GA, Koeva YA, Atanassova PK, Andonov VN, Trichkova VA. Serotonin-producing cells in human gastric mucosa--immunohistochemical and electron microscopic study. Folia Med (Plovdiv). 2010 Apr-Jun;52(2):31-7. PubMed PMID: 20836394.

Pévet P, Agez L, Bothorel B, Saboureau M, Gauer F, Laurent V, Masson-Pévet M. Melatonin in the multi-oscillatory mammalian circadian world. Chronobiol Int. 2006;23(1-2):39-51. Review. PubMed PMID: 16687278.

Puigjaner J, Fábrega J, de Diego I, Subirada F, Durany O, Rivero-Urgell M. Two circadian infant formulas produce differential cerebellum gene expression in lactating rat neonates. *Genes Nutr.,* 2007 Oct;2(1):129-31. doi:10.1007/s12263-007-0037-2. PubMed PMID: 18850161; PubMed Central PMCID:PMC2474909.

Reinberg A, Ashkenazi I. Concepts in human biological rhythms. Dialogues Clin Neurosci. 2003 Dec;5(4):327-42. PubMed PMID: 22033796; PubMed Central PMCID:PMC3181781.

Revollo JR, Grimm AA, Imai S. The regulation of nicotinamide adenine dinucleotide biosynthesis by Nampt/PBEF/visfatin in mammals. Curr Opin Gastroenterol. 2007 Mar;23(2):164-70. Review. PubMed PMID: 17268245.

Rhoads RE, Grudzien-Nogalska E. Translational regulation of milk protein synthesis at secretory activation. J Mammary Gland Biol Neoplasia. 2007 Dec;12(4):283-92. Epub 2007 Nov 17. Review. PubMed PMID: 18026822.

Richard DM, Dawes MA, Mathias CW, Acheson A, Hill-Kapturczak N, Dougherty DM. L-Tryptophan: Basic Metabolic Functions, Behavioral Research and Therapeutic Indications. Int J Tryptophan Res. 2009 Mar

23;2:45-60. PubMed PMID: 20651948; PubMed Central PMCID: PMC2908021.

Rivero U M., Santamaría O A, Rodríguez-Palmero S M. Nutrición Hospitalaria, 2005, 20(2).pp135-146.

Rudloff S, Kunz C. Protein and nonprotein nitrogen components in human milk, bovine milk, and infant formula: quantitative and qualitative aspects in infant nutrition. *J. Pediatr. Gastroenterol. Nutr.,* 1997 Mar;24(3):328-44. Review. PubMed PMID: 9138181.

Ruddick JP, Evans AK, Nutt DJ, Lightman SL, Rook GA, Lowry CA. Tryptophan metabolism in the central nervous system: medical implications. *Expert. Rev. Mol. Med.,* 2006 Aug 31;8(20):1-27. Review. PubMed PMID: 16942634.

Sainio EL, Pulkki K, Young SN. L-Tryptophan: Biochemical, nutritional and pharmacological aspects. Amino Acids. 1996 Mar;10(1):21-47. doi:10.1007/BF00806091. PubMed PMID: 24178430.

Sakamoto K, Oishi K, Nagase T, Miyazaki K, Ishida N. Circadian expression of clock genes during ontogeny in the rat heart. *Neuroreport,* 2002 Jul 19;13(10):1239-42. PubMed PMID: 12151777.

Sampson DA, Hunsaker HA, Jansen GR. Dietary protein quality, protein quantity and food intake: effects on lactation and on protein synthesis and tissue composition in mammary tissue and liver in rats. *J. Nutr.,* 1986 Mar;116(3):365-75. PubMed PMID: 3950763.

Sánchez S, Sánchez C, Paredes SD, Cubero J, Rodríguez AB, Barriga C. Circadian variations of serotonin in plasma and different brain regions of rats. *Mol. Cell Biochem.,* 2008 Oct;317(1-2):105-11. doi: 10.1007/s11010-008-9836-z. Epub 2008 Jun 19. PubMed PMID: 18563534.

Sanger GJ. 5-hydroxytryptamine and the gastrointestinal tract: where next? *Trends Pharmacol. Sci.,* 2008 Sep;29(9):465-71. doi: 10.1016 /j.tips.2008.06.008. Review. PubMed PMID: 19086255.

Sarwar G. Influence of tryptophan supplementation of soy-based infant formulas on protein quality and on blood and brain tryptophan and brain serotonin in the rat model. *Plant Foods Hum. Nutr.,* 2001;56(3):275-84. PubMed PMID: 11442227.

Shimomura H, Moriya T, Sudo M, Wakamatsu H, Akiyama M, Miyake Y, Shibata S. Differential daily expression of Per1 and Per2 mRNA in the suprachiasmatic nucleus of fetal and early postnatal mice. *Eur. J. Neurosci.,* 2001 Feb;13(4):687-93. PubMed PMID: 11207804.

Sidransky H. Tryptophan: biochemical and health implications. CRC Press, 2001.

Sládek M, Jindráková Z, Bendová Z, Sumová A. Postnatal ontogenesis of the circadian clock within the rat liver. *Am. J. Physiol. Regul. Integr Comp. Physiol.,* 2007 Mar;292(3):R1224-9. Epub 2006 Nov 9. PubMed PMID: 17095653.

Steinberg LA, O'Connell NC, Hatch TF, Picciano MF, Birch LL. Tryptophan intake influences infants' sleep latency. *J. Nutr.,* 1992 Sep;122(9):1781-91. PubMed PMID: 1512627.

Steinberg LA, O'Connell NC, Hatch TF, Picciano MF, Birch LL. Tryptophan intake influences infants' sleep latency. *J. Nutr.,* 1992 Sep;122(9):1781-91. PubMed PMID: 1512627.

Sumová A, Bendová Z, Sládek M, El-Hennamy R, Laurinová K, Jindráková Z, Illnerová H. Setting the biological time in central and peripheral clocks during ontogenesis. *FEBS Lett.,* 2006 May 22;580(12):2836-42. Epub 2006 Mar 20. Review. PubMed PMID: 16563387.

Sumová A, Bendová Z, Sládek M, Kovácikovà Z, El-Hennamy R, Laurinová K, Illnerová H. The rat circadian clockwork and its photoperiodic entrainment during development. *Chronobiol. Int.,* 2006;23(1-2):237-43. PubMed PMID: 16687297.

Sumová A et al. Clock gene daily profiles and their phase relationship in the rat suprachiasmatic nucleus are affected by photoperiod. Journal of biological rhythms, v. 18, n. 2, p. 134-144, 2003.Sumová A, Jác M, Sládek M, Sauman I, Illnerová H. Clock gene daily profiles and their phase relationship in the rat suprachiasmatic nucleus are affected by photoperiod. *J. Biol. Rhythms.,* 2003 Apr;18(2):134-44. PubMed PMID: 12693868.

Trindade-filho E, Vasconcelos M, Carlos Augusto Carvalho de; GUEDES, Rubem Carlos Araújo. Acute tryptophan administration impairs cortical spreading depression propagation in REM sleep deprived and non-deprived adult rats. Psychology & Neuroscience, v. 2, n. 2, p. 235-241, 2009.

Tsopmo A, Diehl-Jones BW, Aluko RE, Kitts DD, Elisia I, Friel JK. Tryptophan released from mother's milk has antioxidant properties. Pediatr Res. 2009 Dec;66(6):614-8. doi: 10.1203/PDR.0b013e318 1be9e7e. PubMed PMID: 19730158.

Uçeyler N, Schütt M, Palm F, Vogel C, Meier M, Schmitt A, Lesch KP, Mössner R, Sommer C. Lack of the serotonin transporter in mice reduces locomotor activity and leads to gender-dependent late onset obesity. *Int. J. Obes* (Lond). 2010 Apr;34(4):701-11. doi: 10.1038/ijo.2009.289. Epub 2010 Jan 19. PubMed PMID: 20084070.

Vallone D, Lahiri K, Dickmeis T, Foulkes NS. Start the clock! Circadian rhythms and development. Dev Dyn. 2007 Jan;236(1):142-55. Review. PubMed PMID: 17075872.

Van de Kar LD, Rittenhouse PA, Li Q, Levy AD. Serotonergic regulation of renin and prolactin secretion. *Behav. Brain Res.,* 1996;73(1-2):203-8. Review. PubMed PMID: 8788503.

Varcoe TJ, Boden MJ, Voultsios A, Salkeld MD, Rattanatray L, Kennaway DJ. Characterisation of the maternal response to chronic phase shifts during gestation in the rat: implications for fetal metabolic programming. PLoS One. 2013;8(1):e53800. doi: 10.1371/journal.pone.0053800. Epub 2013 Jan 14. PubMed PMID: 23342007; PubMed Central PMCID: PMC3544759.

Vucetic Z, Kimmel J, Totoki K, Hollenbeck E, Reyes TM. Maternal high-fat diet alters methylation and gene expression of dopamine and opioid-related genes. Endocrinology. 2010 Oct;151(10):4756-64. doi: 10.1210/en.2010-0505. Epub 2010 Aug 4. PubMed PMID: 20685869; PubMed Central PMCID: PMC2946145.

Vucetic Z, Totoki K, Schoch H, Whitaker KW, Hill-Smith T, Lucki I, Reyes TM. Early life protein restriction alters dopamine circuitry. Neuroscience. 2010 Jun 30;168(2):359-70. doi: 10.1016/j.neuroscience.2010.04.010. Epub 2010 Apr 13. PubMed PMID: 20394806; PubMed Central PMCID: PMC2873068.

Wang XS, Armstrong ME, Cairns BJ, Key TJ, Travis RC. Shift work and chronic disease: the epidemiological evidence. Occup Med (Lond). 2011 Mar;61(2):78-89. doi: 10.1093/occmed/kqr001. Review. PubMed PMID: 21355031; PubMed Central PMCID: PMC3045028.

Wideman CH, Murphy HM. Constant light induces alterations in melatonin levels, food intake, feed efficiency, visceral adiposity, and circadian rhythms in rats. *Nutr. Neurosci.,* 2009 Oct;12(5):233-40. doi:10.1179 /147683009X423436. PubMed PMID:19761654.

Winokur A, Lindberg ND, Lucki I, Phillips J, Amsterdam JD. Hormonal and behavioral effects associated with intravenous L-tryptophan administration. Psychopharmacology (Berl). 1986;88(2):213-9. PubMed PMID: 3081933.

Yamaoka I, Nakayama M, Miki T, Yokoyama T, Takeuchi Y. Dietary protein modulates circadian changes in core body temperature and metabolic rate in rats. *J. Physiol. Sci.,* 2008 Feb;58(1):75-81. doi: 10.2170 /physiolsci.RP006707. Epub 2008 Feb 2. PubMed PMID: 18237456.

Yamazaki S, Yoshikawa T, Biscoe EW, Numano R, Gallaspy LM, Soulsby S, Papadimas E, Pezuk P, Doyle SE, Tei H, Sakaki Y, Block GD, Menaker M. Ontogeny of circadian organization in the rat. *J. Biol. Rhythms,* 2009 Feb;24(1):55-63. doi:10.1177/0748730408328438. PubMed PMID: 19150929; PubMed Central PMCID:PMC2665126.

Yogman MW, Zeisel SH. Nutrients, neurotransmitters and infant behavior. *Am. J. Clin. Nutr.,* 1985 Aug;42(2):352-60. Review. PubMed PMID: 2862787.

Yonekura R et al. The effects of dietary tryptophan metabolism on the “vigor” and “confusion” factor of mood. In: International Congress Series. Elsevier, 2007. p. 180-183.

Young ME. The circadian clock within the heart: potential influence on myocardial gene expression, metabolism, and function. Am J Physiol *Heart Circ. Physiol.,* 2006 Jan;290(1):H1-16. Review. PubMed PMID: 16373589.

Young VR, Marchini JS. Mechanisms and nutritional significance of metabolic responses to altered intakes of protein and amino acids, with reference to nutritional adaptation in humans. *Am. J. Clin. Nutr.,* 1990 Feb;51(2):270-89. Review. PubMed PMID: 2407099.

Yu PL, Fujimura M, Okumiya K, Kinoshita M, Hasegawa H, Fujimiya M. Immunohistochemical localization of tryptophan hydroxylase in the human and rat gastrointestinal tracts. *J. Comp. Neurol.,* 1999 Sep 6;411(4):654-65. PubMed PMID:10421874.

Zhang T, Berrocal JG, Frizzell KM, Gamble MJ, DuMond ME, Krishnakumar R, Yang T, Sauve AA, Kraus WL. Enzymes in the NAD+ salvage pathway regulate SIRT1 activity at target gene promoters. J Biol Chem. 2009 Jul 24;284(30):20408-17. doi: 10.1074/jbc.M109.016469. Epub 2009 May 28. PubMed PMID: 19478080; PubMed Central PMCID: PMC2740465.

Zvonic S, Ptitsyn AA, Conrad SA, Scott LK, Floyd ZE, Kilroy G, Wu X, Goh BC, Mynatt RL, Gimble JM. Characterization of peripheral circadian clocks in adipose tissues. *Diabetes,* 2006 Apr;55(4):962-70. PubMed PMID: 16567517.

In: New Developments in Tryptophan Research ISBN: 978-1-63483-025-6
Editor: Victoria Hayes

Chapter 4

TRYPTOPHANASE ENANTIOSELECTIVITY TRANSFORMING D-SERINE FROM INHIBITOR TO ACTIVE SUBSTRATE IN THE PRESENCE OF TRIAMMONIUM PHOSPHATE

Akihiko Shimada* and Ikumi Otsuka
Sustainable Environment Studies, Graduate School of Life and Environmental Sciences, University of Tsukuba, Tsukuba, Japan

ABSTRACT

A historical riddle about chiral homogeneity, the origin of L-dominant amino acids in biological world, has intrigued scientists for many years, though there has been no general consensus as yet to it. Life could not been born without homochirality, that is to say, no homochirality – no life. However, we cannot imagine at all when and how the homochirality was established on early earth. Amino acids (also sugars) are stereochemically too labile to retain their homochiral state in natural environment because racemization pressure is overwhelmingly strong, especially in aqueous solution. An elaborate mechanism is necessary to put biological world into homochirally-stable state. The homochirality of contemporary biological world is fully guaranteed by enzyme enantioselectivity. Since the enzyme enantioselectivity has

* Corresponding author: Akihiko Shimada. E-mail: ashimada@envr.tsukuba.ac.jp.

remained unclear, a deeper understanding of its mechanism is required to unravel all the riddle of the origin of homochirality.

We have so far researched a mechanism of tryptophanase enantioselectivity, with the aim of answering its problem. Tryptophanase degrades L-tryptophan by β-elimination reaction, and also synthesizes L-tryptophan from L-serine and indole by β-replacement reaction by use of its absolute enantioselectivity. On the other hand, it is perfectly inert to D-tryptophan and D-serine at all. However, our previous studies reported that tryptophanase become active on D-tryptophan and D-serine in the presence of diammoniumhydrogen and triammonium phosphate. These salts make the enantioselectivity of tryptophanase flexible. The present research provides triammonium phosphate is the most effective salt among several salts to make the enantioselectivity flexible. A metabolic pathway of L-tryptophan synthesis from D-serine were determined to elucidate how ammonium phosphates produced the flexible enantioselectivity, indicating it was the most significant to form an aldimine bond between amino group of D-serine and pyridoxal 5'-phosphate at the first step of the reaction. Investigation of spectroscopically-based conformational change provided the flexible enantioselectivity was emerged by reversible small conformational change in the presence of triammonium phosphate. We describe these new research findings about L-tryptophan synthesis from D-serine, along with approach to tryptophan research from a new angle.

Keywords: Enantioselectivity, tryptophanase, triammonium phosphate, D-serine, L-tryptophan synthesis, origin of homochirality

INTRODUCTION

The so-called chiral breaking problem has intrigued scientists for many years. There have been many reports on natural-occurring chiral breaking due to asymmetric decomposition with hydrolysis or circular polarized light [1]. However, there has been no general consensus as yet to the origin of homochirality. One of those reasons is responsible for enantiomerically labile amino acids. A half-life of biological amino acid racemization is 10^3 years for phenylalanine or 10^5 years for glutamic acid at 298 K in aqueous solution [2]. On the other hand, the duration needed for the emergence of life is more than seven million years [3]. In the opinion of some scientists, this period is much too short for the evolution of life. Consequently they suggest an extraterrestrial evolution of life and assume the delivery of sperm to Earth *via* the atmosphere [4]. This panspermia theory, published by its most famous protagonist

Arrhenius [5], obviously just shifts the problem from Earth to space, and seems to give rise to more problems than it solves. Therefore, we neglect it, and again consider a process of the amino acid chiral homogeneity during the short period. Even if one-handed amino acids had just happened to appear on earth with only a slight enantiomeric excess before the creation of life, any slight excess would have been quickly extinguished by racemization in water [6]. Life might have not existed if enantiomers had been heterogeneous at an earlier stage of the creation of life. Even now, racemization pressure is still overwhelmingly strong in natural environment, especially in aqueous solution [7]. Amino acids are stereochemically too labile to keep their enantiomeric type homochiral in the natural environment. It is thought that life had to create a new elaborate mechanism to generate homochirality in early stage. That process of birth of homochirality has remained unknown, but the present biological world has been in a homochirally-stable state. For example, amino acids, which can be regarded as components of all polypeptides, must be homogeneous to assemble a regularly structured polypeptide. If a polypeptide is mixed with L- and D- amino acids, it is no longer possible to create an ultra-precise structure because the racemic mixture of enantiomers cannot elongate the polypeptide chain [8]. In this context, life itself can be regarded as an ultimately homoorganized structure made up of homochiral biomolecules. Thus, early life's first strategy is to homogenize racemic amino acids to use as the components of protein or enzyme. Perhaps some sophisticated contrivance is required to prepare homochiral amino acid molecules enough to synthesize polypeptides. At least a systematic mechanism with automatically developing chiral homogeneity will be proposed for it, and additionally it must has a very complicated mechanism that selects and delivers the right enantiomeric type to the right place while at the same time excluding the opposite enantiomer in order to synthesize regularly structured macromolecules [9]. By the time catalytically active early polypeptides, i.e., primitive enzymes, were present, the switch to the exclusive use of L-amino acids might already have been made. If early polypeptides were thus synthesized by spontaneous abiotic processes in a primitive racemic environment, the mechanism of homochirality may have already been incorporated into them, and their descendants may be traced to present enzymes. Such a mechanism is involved in extant enzyme enantioselectivity. Enantioselectivity must be absolutely stable because unstable one connects to death of life. We can say life does not exist without homochirality, i.e., no homochirality - no life [10]. This principle is so fundamental that it is inseparably associated with the origins of life. The substrate specificity of many enzymes on L-amino acids and their catalytic mechanisms are studied with the corresponding substrates, while the mechanism of their enantioselectivity is only poorly understood. A deeper understanding of it is necessary to unravel the mystery of the origin of homochirality.

A stable supply of homochiral molecules, which is necessary to synthesize biomacromolecule with highly organized stereostructure such as protein, is achieved by enzyme enantioselectivity that functions to select the right enantiomer; and therefore the stability of this enantioselectivity is essential for sustaining vital activity [11]. However, many reports regarding physiologically active D-amino acids have been provided for the last dozen years or so [12]. It indicate the enantioselectivity is flexible contrary to conventional wisdom. We have studied about the enantioselectivity of tryptophanase (TPase), with the aim of acquiring a better understanding of the enantioselectivity. TPase is an enzyme that degrades L-tryptophan (L-Trp) into indole, pyruvate and ammonia, or synthesizes L-Trp from L-serine (L-Ser) and indole. It is also known as an enzyme that has a very wide substrate specificity for L-Trp derivatives and various β-substituted L-amino acids on the one hand, but at the same time has an extremely tight enantioselectivity [13]. Because of this absolute enantioselectivity, tryptophanase has no activity on D-tryptophan (D-Trp) at all. However, our previous studies have shown that TPase becomes active towards the D-Trp in highly concentrated diammoniumhydrogen phosphate (DAP) solution [14]. TPase enantioselectivity is very flexible in the presence of DAP [15]. D-Trp was degraded into indole, pyruvate and ammonia through β-elimination after D-Trp formed an aldimine bond with pyridoxal 5'-phosphate (PLP) [16]. The formation of the external aldimine bond between the D-enantiomer and PLP was the first step for D-enantiomer to function as active substrate. However, even if C^{α}-NH_2 succeeded in the formation of the external aldimine bond with PLP, subsequent reactions did not always proceed automatically. Several inhibitors with amino groups, for example, L-phenylalanine or L-methionine, can also form the external aldimine bond with PLP, but they can never become substrates for TPase [17]. It indicates that the forming of the aldimine bond is a necessary prerequisite for the flexible enantioselectivity but it is not enough to catalyze D-enantiomer. A necessary and sufficient condition is the existence of DAP. Previous reports showed that TPase underwent a small reversible conformational change in the presence of DAP [18]. Perhaps factor such as conformational change may be required as well.

We think this small conformational change possibly modifies the enantioselectivity of TPase so that D-enantiomer can be activated *via* the external aldimine bond formation. Inhibitors, which structurally resemble their enzyme substrate but either do not react or react only very slowly compared to the substrate, are commonly used to probe the conformational nature of a substrate-binding site to elucidate the enzymes' catalytic mechanisms. Therefore, when a tryptophan-analogous inhibitor is used for kinetic analysis, it is powerful enough to understand what happens at the active site of TPase with a conformational change [19]. D-Ser is an inhibitor of TPase, and thereby has no activity on TPase at all. When we made a kinetic analysis of TPase to study a β-substitution mechanism of the ping-pong type in L-Trp synthesis from L-Ser with using D-Ser inhibitor, we found D-Ser transformed from an inhibitor into an active substrate in the presence of DAP or triammonium phosphate (TAP) [20]. TPase produces pyruvate from D-Ser *via* a β-elimination reaction pathway in the presence of TAP, but when indole is present, it can synthesize L-Trp *via* β-replacement without pyruvate production. Although D-Ser is excluded from the original metabolism of TPase in the absence of TAP, it can be accepted as an active substrate through a small stereostructual change due to TAP. We newly found D-Ser turns into active substrate in the presence of TAP. This synthesis is especially more attractive than degradation because birth of life is seen as the results of integrated myriad synthetic reactions. The present study aims to describe TPase-catalyzed L-tryptophan synthesis from D-Ser due to flexible enantioselectivity in the presence of TAP.

D-SERINE AS COMPETITIVE INHIBITOR

Generally speaking, D-amino acid works as inactive or inhibitor against enzyme. D-Ser is also inactive on TPase. Kinetic analysis is effective to probe the interaction between D-Ser and TPase. We analyzed D-Ser behavior towards TPase in terms of enzyme kinetics. For kinetics, reaction was conducted at 50°C for 4 hr in Dry Thermo Unit DTU-1B (Taitec, Tokyo) with 0, 0.1, 0.2, 0.3, 0.6, or 1 M of D-Ser added in reaction mixtures that were prepared by combining the required L-Ser (10-100 mM) with fixed apotryptophanase (apoTPase, 0.46 μM), indole (5.4 mM) and PLP concentrations (0.4 mM) in 500 μL of 0.1 M potassium phosphate buffer (PB), adjusted to a pH of 8.3. After 4hr, the reaction mixture was cool on ice to stop reaction. 200 μL of the reaction mixture was filtered through Amicon Ultra Centrifugal Filters cutting more than MW 3000 (Millipore, Corrigtwohill, Co. Cork, Ireland), immediately injected on a Crownpack CR (+) Resolution

HPLC resolution column (DAICEL, Osaka, Japan) equilibrated with the degassed eluent through a Degasser, and eluted at a flow rate of 0.9 mL/min at room temperature with an HPLC pump. L-Trp synthesis from L-Ser was determined measuring with a CD detector (CD-1595, JASCO, Tokyo, Japan). Ellipticity [θ] was monitored at $\lambda = 230$ nm with the CD detector, represented in units of mdeg. Ellipticity of the eluent was set up to zero at a wavelength of $\lambda = 230$ nm, so the peaks of L-Trp appeared on the positive side. For kinetic analysis, TPase activity on L-Trp synthesis was assayed in triplicate at the same substrate concentration, averaged. Lineweaver-Burk plots were drawn based on the least squares method. The reciprocal of rate constant (ordinate) versus the reciprocal of substrate concentration (abscissa) was plotted for each concentration of D-Ser in Figure 1. Kinetic constants were calculated from these double-reciprocal plots, and also an inhibition constant K_I of D-Ser was determined from their slopes. K_m was 44 mM, V_m 2.2 $\times 10^{-4}$ mM/s, k_{cat} 0.48/s and K_I 419 mM. Figure 1 depicted a typical competitive inhibition plot. D-Ser is a competitive inhibitor against TPase in L-Trp synthesis reaction from L-Ser and indole. Competitive inhibitor usually shares a highly structural resemblance with its substrate. D-Ser stereostructure fits into the active site structure of TPase, promptly binding there. However, D-Ser cannot be involved in L-Trp synthetic reaction. Perhaps D-Ser cannot form aldimine-bond with PLP at the first step of L-Trp synthesis. The presence of TAP promotes a small conformational change to TPase to fit D-Ser partly into the catalytic site of the active site and help D-Ser to form the external aldimine bond with PLP. This subtle steric interaction, whose detail is still being investigated, are very important to transform D-Ser from inhibitor into active substrate.

Effects of Various Salts on L-Trp Synthesis from D-Ser

Concentrate salt solution such as TAP is necessary to conduct activity on D-Ser. It is be of enzymological significance to investigate whether or not the activity occurs in other salt solutions. Thus we investigated about other various salt species to bring about the activity.

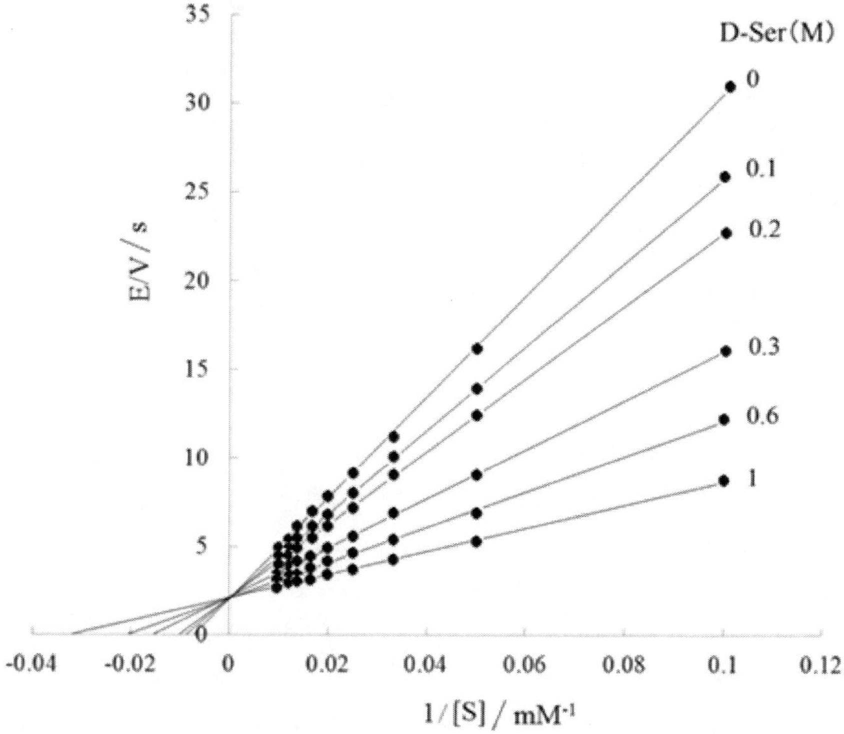

Figure 1. The reciprocal plots of rate constant versus substrate concentration for each concentration of D-Ser in L-Trp synthesis from L-Ser.

The salts tested are as follows; 9 Good's buffers (TAPSO, Tricine, HEPES, Bicine, BES, CHES, MES, TAPS, TES), ammonium sulfate, ammonium chloride, Tris(hydroxylmethyl)aminomethane (Tris-HCl), ammonium dihydrogenphosphate (MAP), DAP, TAP. Saturated solution of each salt was diluted to the required saturation concentration by adding 0.1 M PB of pH 8.3. Composition of each reaction mixture was as follows; 20% saturation concentration of each salt, 0.4 mM PLP, 5.4 mM indole, 100 mM D-Ser and 0.46 μM apoTPase in L-Trp synthesis from D-Ser; or 50% saturation concentration of each salt, 1.1 mM PLP, 245 μM D-Trp and 0.23 μM apoTPase in D-Trp degradation. Reaction temperature and time was 60 °C and 6 hr in L-Trp synthesis from D-Ser, or 60°C and 4 hr in D-Trp degradation. The activity was determined as described above or in reference [15]. Table 1 showed the activity on L-Trp synthesis from D-Ser and D-Trp degradation in each salt solution.

Table 1. Effects of various on activity on D-Ser

Salt species	Activity on L-Trp synthesis from D-Ser	Activity on D-Trp degradation
TAPSO	–	–
Tricine	–	±
HEPES	–	±
Bicine	–	–
BES	–	–
CHES	–	–
MES	–	±
TAPS	–	±
TES	–	–
ammonium sulfate	–	+
ammonium chloride	–	–
Tris-HCl	–	–
MAP	–	–
DAP	++	++
TAP	+++	+++

–: no activity, ±: minimum detectable activity, +: detectable activity, ++: strong activity, +++: the strongest activity.

Most salts did not provide the activity on these D-amino acids. However, several salts made it positive. Tricine, HEPES, MES and TAPS of Good buffers in addition to ammonium sulfate had trivial activity in D-Trp degradation, while they had no activity on L-Trp synthesis from D-Ser. This shows enzyme enantioselectivity is more flexible in degradation than in synthesis because degradative chemical reaction is energy-releasing process to break down complex molecules into smaller units. In other words, it indicates anabolic metabolism demands very tight enantioselectivity to build a large refined polymer made of many small monomers such as amino acids. Ammonium phosphates have 3 species of MAP, DAP and TAP. MAP irreversibly inactivated to denature TPase, but DAP and TAP provided the activity on D-amino acids in Table 1. TAP had 2.3-fold stronger activity on D-Ser than DAP. When DAP or TAP is removed from reaction mixture, TPase is reversibly restored to its original state. The activity was assayed in the range of 10 to 40% of saturation concentration of TAP or at the range of 30 to 80°C of reaction temperature to determine optimal conditions of L-Trp synthetic reaction from D-Ser in the presence of TAP. Figure 2 and 3 show the activity changing with TAP saturation and reaction temperature at a fixed reaction temperature of 60°C and saturation concentration of 30%, respectively. An optimal TAP concentration and reaction temperature were 30% saturation and 55°C, respectively. Maximal L-Trp synthesis from D-Ser was 3.2 mM. On the other hand, the activity on L-Trp synthesis from L-Ser also was examined in the absence of TAP to compare with the activity on D-Ser. Its maximal activity was 3.15 mM at a reaction temperature of 50°C. When the activity on D-Ser and L-Ser was compared in Figure 3, the former was 10.2% of the latter.

NO AFFECT OF TAP ON D-SER RACEMIZATION

We are concerned as to whether or not D-Ser racemized while D-Ser is exposed to 55°C for 6 hr in 30% saturation TAP solution. Prior to experiment, perchloric acid was added dropwise to a few liters of distilled water with vigorous stirring at room temperature until pH decreased to 1.0, and used at 0°C as an eluent for HPLC.

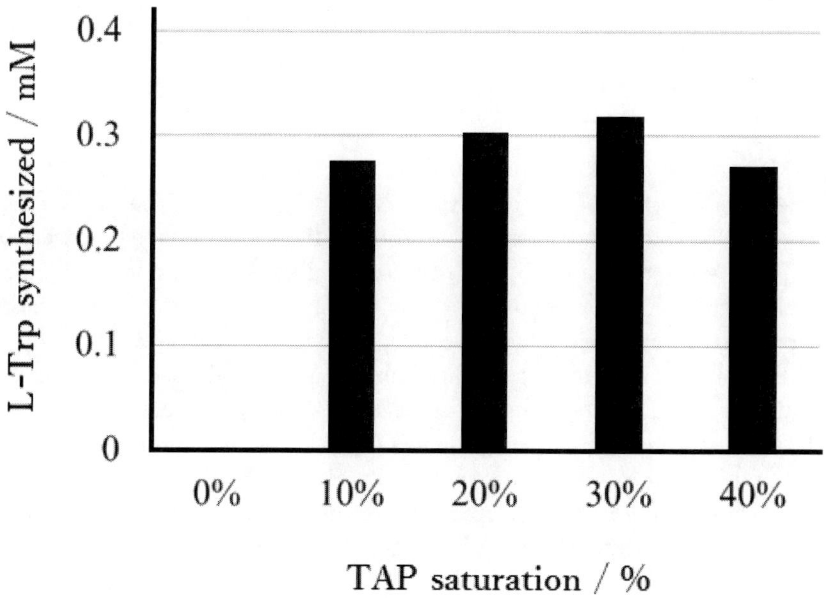

Figure 2. Activity on L-Trp synthesis from D-Ser depending on TAP saturation concentrations.

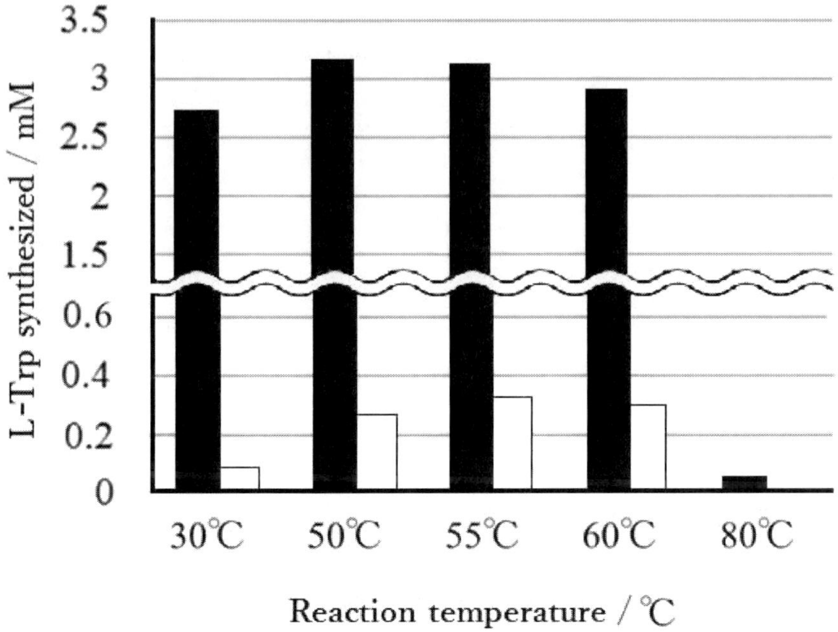

■: L-Ser, □: D-Ser.

Figure 3. Activity on L-Trp synthesis from L-Ser or D-Ser depending on reaction temperature.

TAP solution of 30% saturation including 100 mM D-Ser, 0.4 mM PLP and 5.4 mM indole was exposed to 55°C for 6 hr. After 6hr, D-Ser was resolved at 0°C at a flow rate of 0.1 mL/min in the same method as described above. UV absorbance and ellipticity [θ] was simultaneously monitored at λ = 230 nm with the CD detector. Elliptical peaks of D-serine and L-serine appeared on the negative and positive side, respectively. 100 mM DL-serine in 0.1 M PB solution was first resolved for control in UV and CD chromatograms of Figure 4a. In UV chromatogram, a peak of D-Ser preceded one of L-Ser, and DL-Ser was almost resolved. D- and L-Ser were eluted at retention time of 20 min and 23 min, respectively. On the other hand, D-Ser perfectly appeared on the negative side, and L-serine on the positive side in CD chromatogram. This resolution LC system was adequately beneficial to D-Ser racemization analysis. Subsequently 100 mM D-serine was exposed to 55°C for 6 hr in 30% saturation TAP, immediately resolved in Figure 4b. If D-Ser is racemized into L-Ser, any peak would be found on the positive side at L-Ser retention time of 23 min in both of UV and CD chromatograms.

However, no peak was seen in either chromatogram (Figure 4b). This indicates that D-Ser is not racemized at all even if it is exposed to TAP at enzymologically high temperature of 55°C. D-Ser works on TPase as a substrate to synthesize L-tryptophan.

REACTION PATHWAY OF L-TRP SYNTHSIS FROM D-SER

Three possible reaction pathways can be proposed for L-Trp synthesis from D-Ser. They are itemized below.

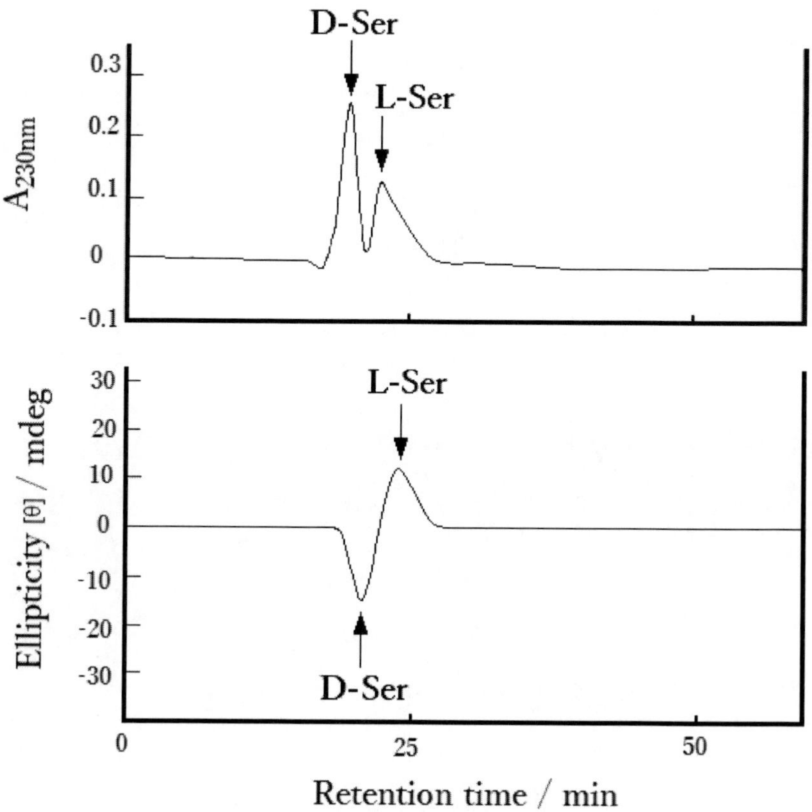

Upper: UV chromatogram, lower: CD chromatogram.

Figure 4a. UV and CD resolution chromatograms of D-Ser in 0.1 M potassium phosphate buffer solution.

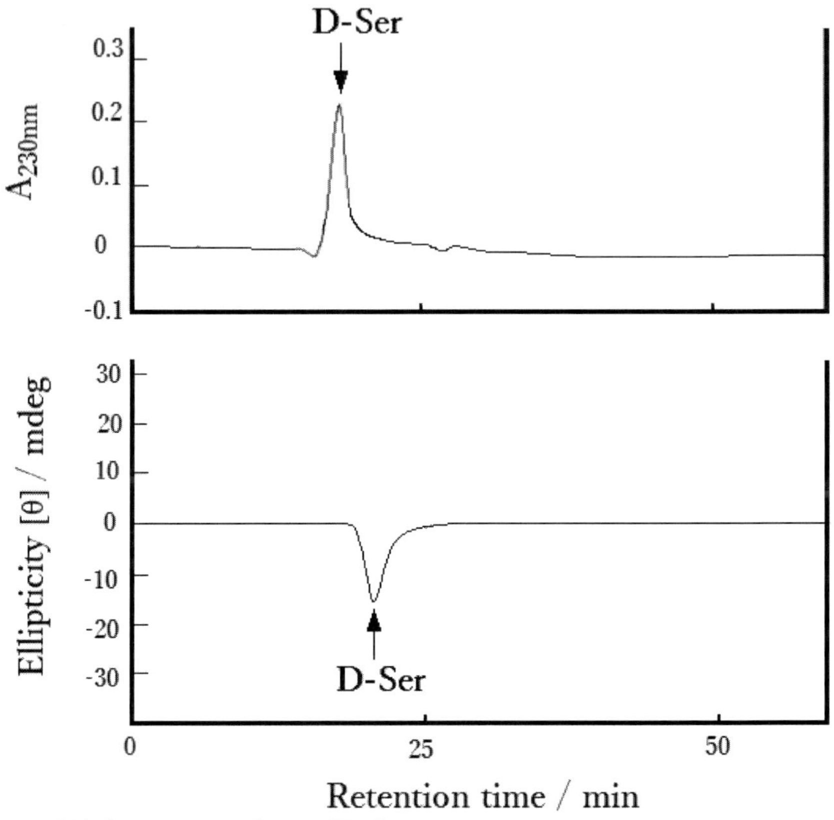

Upper: UV chromatogram, lower: CD chromatogram.

Figure 4b. UV and CD resolution chromatograms of D-Ser exposed to 55 °C for 6 hr in 30% saturation TAP solution.

1. Racemization of D-Serine

Racemization of D-serine into L-serine is first suspected as a cause of the emergence of L-Trp synthetic activity on D-Ser in the presence of TAP.

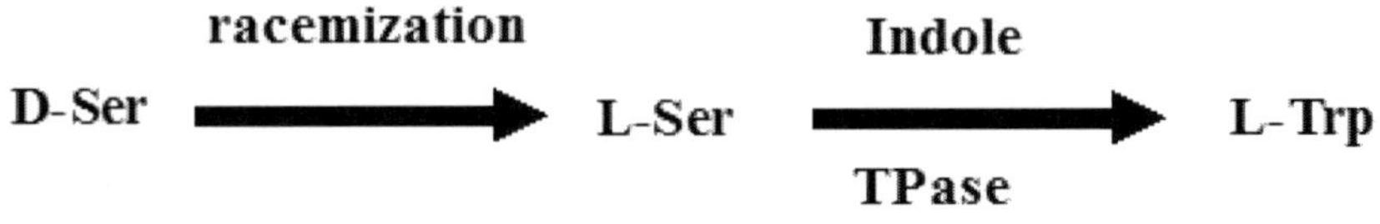

2. α, β-Elimination Route

D-Ser forms TPase-bound aminoacrylate intermediate in the presence of TAP, and subsequently the intermediate produces pyruvate *via* α, β-elimination route. The pyruvate is synthesized into L-tryptophan together with indole and ammonia.

$$\text{D-Ser} \xrightarrow[\text{TPase}]{\alpha,\ \beta\text{-elimination}} \text{Pyruvic acid} \xrightarrow[\text{TPase}]{\text{Indole, } NH_3} \text{L-Trp}$$

3. β-Replacement Route

The TPase-bound aminoacrylate intermediate does not follow α, β-elimination but β-replacement route. The intermediate produces L-tryptophan *via* a TPase-bound quinonoid intermediate. This β-replacement reaction is the same reaction pathway as tryptophan synthase, though no structural similarity is detected between the two enzymes [21].

$$\text{D-Ser} + \text{Indole} \xrightarrow[\text{TPase}]{\beta\text{-replacement}} \text{L-Trp}$$

The first item 1 suggests a possibility of D-Ser racemization when D-Ser is exposed to concentrate TAP at high temperature for hours. No racemization of D-Ser was previously determined in Figure 4a and b. D-Ser is incorporated directly into TPase active site as a substrate, and immediately binding there to produce TPase-bound D-serine, which is converts to L-Trp *via* TPase-bound aminoacrylate intermediate. The second item 2 thinks of the pyruvate that is degraded from D-Ser by α, β-elimination reaction. TPase can degrade D-Ser into pyruvate in the presence of TAP, and also it is well known that indole is a strong inhibitor of serine deamination, or β-elimination leading to pyruvate formation from L-serine. Thereby, we investigated how indole inhibited D-Ser deamination in the presence of TAP. TPase ceased serine deamination when an indole concentration was over 5.0 mM [20].

Since we used a 5.4 mM of indole concentration in L-tryptophan synthesis from D-Ser, TPase fell into a condition of the perfect cessation of the pyruvate formation. Our result provided TPase produced L-Trp from D-Ser without pyruvate production. It indicated that L-tryptophan was synthesized from D-serine without passage of β-elimination. We still suspected the involvement of pyruvate in L-Trp synthesis from D-Ser. An additional experiment analyzed L-Trp synthesis in the presence of 0.7 mM pyruvate, which corresponded to the pyruvate production from 100 mM D-Ser *via* α, β-elimination at 55°C and for 6h, showing no L-Trp production. These results indicate pyruvate is not involved in L-Trp synthesis from D-Ser. We can thus confirm a reaction pathway of the last item 3, in which L-tryptophan synthesis from D-serine takes place *via* the β-replacement reaction pathway.

TRANSFORMATION OF D-SER FROM INHIBITOR INTO ACTIVE SUBSTRATE

Many enzymes are known to bind both enantiomeric types of a substrate. When catalytic reaction is restricted to one enantiomer, the other enantiomer, although bound to the active site, is inert or it functions as an inhibitor [22, 23]. Generally speaking, it is impossible to transform a D-amino acid into an active substrate. Additionally, synthetic reaction should especially demand more absolute stereospecificity than degradative reaction, because the former must recreate an exact replica of life with ultraprecise higher-order architecture. It is hard to find versatile enantioselectivity in the synthesis reaction. TPase is so flexible that domain shifts can be observed with the holo form found in an open conformation and the apo form in open, wide-open, or closed conformation depending on the nature of different bound ions [24]. TPase produces the broad substrate specificity on many β-substituted L-amino acids owing to the conformationally-flexiblity. However, it only holds true with regard to L type of these amino acids. As far as D-amino acids are concerned, the enantioselectivity of TPase completely rejects and excludes them. In this context it seems strange that TAP makes the enantioselectivity of TPase flexible. As far as we know, there have be seen no previous reports in which an enzyme with tough enantiospecificity has activity on D-amino acid. It is of enzymological significance to grapple with this strange enzyme reaction. Additionally the activity is very attractive to research the origin of homochirality.

It is reasonable to consider that emergence of the flexible enantioselectivity of TPase is attributable to its conformational change. We spectroscopically observed how conformation of TPase changed in TAP solution. We used Bis-ANS (4,4'-dianilino-1,1'-binaphthalene-5,5'-disulfonic acid, $\varepsilon = 23 \times 10^3$ $cm^{-1}M^{-1}$ at 395 nm) as a hydrophobic fluorescent probe, which indicated the existence of solvent-exposed hydrophobic domains in a protein's structure, in fluorophotometry (Hitachi F4500, Tokyo, Japan). Emission from 420 to 650nm was measured at an excitation wavelength of 395 nm. Relative fluorescence intensity was measured in arbitrary units. In a circular dichroism spectrophotometer (J805, Jasco, Tokyo, Japan), spectra were recorded at wavelengths from 200 to 400 nm at room temperature. Five scans were accumulated per spectrum, averaged, and expressed as molar ellipticity in degrees $cm^2 \cdot dmol^{-1}$. The CD spectra were analyzed with a CDPro software package [25].

Since there was the possibility that the small difference of fluorescent spectra might be produced by the difference of polarity of the solvent, we compared the fluorescence spectra of TPase-free TAP and PB solutions to check it prior to measurement. No fluorescence was emitted in either solution, and therefore fluorescent spectrum didn't depend on the polarity. Figure 5 shows fluorescent spectral change of TPase in TAP, DAP, MAP and PB at 55°C. Fluorescence intensity was the highest in MAP, followed by DAP, TAP and PB in order of decreasing. The intensity of TAP and DAP had small difference. Unlike PB, DAP and TAP, a large difference was found in MAP because MAP irreversibly unfolds to denature TPase. The effect of the increasing exposure of hydrophobic amino acids is known to be closely related to aggregation of proteins, and any unfolding undoubtedly leads to the loss of all activity [26]. Since the unfolding of TPase was connected with the increasing exposure of hydrophobic amino acids, the fluorescent intensity increased drastically in response to it. Fluorescence spectral data showed that the conformational change of TPase was much larger in MAP than TAP. The same result was also obtained from circular dichroism spectra of TPase in PB, TAP, DAP and MAP solutions in Figure 6. Circular dichroism spectral changes in the 200-250 nm region among four solutions arise from structural changes of α-helix and β-sheet of TPase because their peaks are in the 205-225 nm region [27]. The CD spectrum of TAP was almost close or similar to that of PB in the 200-250 nm region, but the two CD spectra were not completely same. In fact, slight difference was seen in the wavelength range of 205 – 225 nm. It reflects any subtle conformational change in secondary structure. A CD spectrum of TPase in MAP differed distinctively from those of TAP.

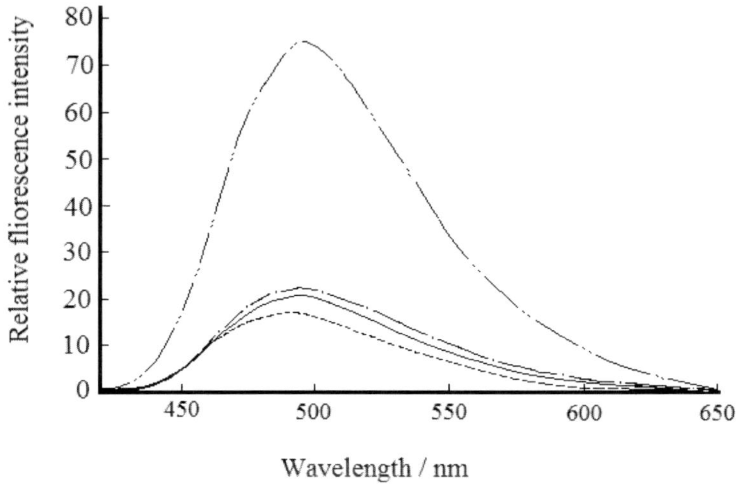

– – –: PB, ——: TAP, — • —: DAP, — • • —: MAP.

Figure 5. Relative fluorescence intensity of bis-ANS-bound TPase in PB, TAP, DAP and MAP solutions at 55 °C.

MAP caused irreversible full loss of the catalytic activity, which probably gave larger structural changes in secondary structure. We previously performed kinetic studies of D-tryptophan degradation reaction [28]. We reported there that TPase returned to the original activity when DAP was removed from the reaction mixture. This means the TPase reverts back to the original conformation. Both fluorescence and circular dichroism spectra provided a very subtle conformational change. TPase can take either closed or open conformations according to variations of the coordination geometry of bound cation [29, 30]. In other words this stereostructurally-flexible TPase is remote cause to give rise to the flexible enantioselectivity through the subtle conformational change.

Speculation over the Advent of L-Amino Acid - Dominant Homochiral Contemporary Biological World

A barrier discriminating between D- and L-amino acid, which serves as the driving force for chiral homogeneity, is very obstinate and headstrong.

Exclusive selection system of L-amino acid can only be built on a solid foundation of absolute enantioselectivity.

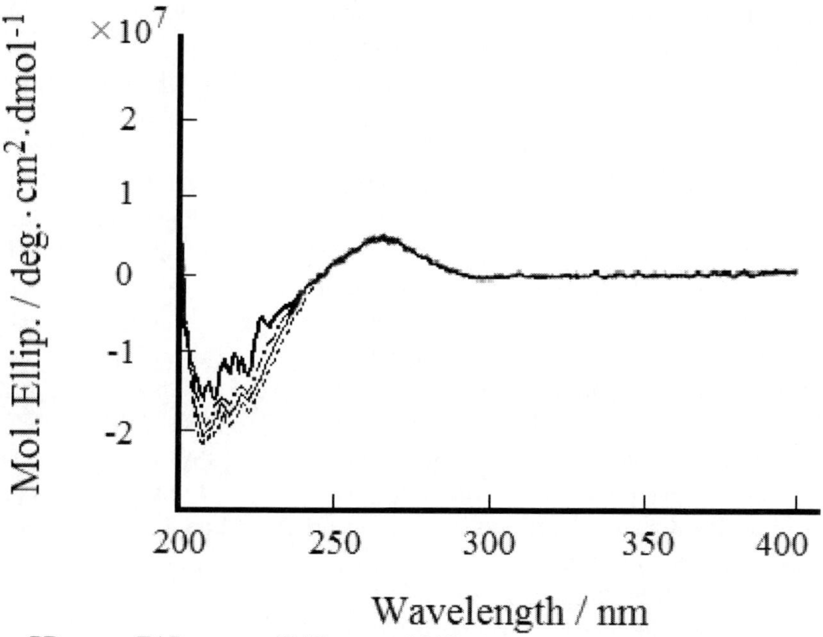

– – –: PB, ——: TAP, — • —: DAP, ▬▬: MAP.

Figure 6. Circular dichroism spectra of TPase in PB, TAP, DAP and MAP solutions at room temperature.

TPase also is one of enzymes with absolute enantioselectivity, whose enantioselectivity reversibly changes to flexible one by small conformational change in concentrated TAP solution [31]. Tyrosine phenol-lyase is an enzyme closely related with TPase. Space structural chemical research on tyrosine phenol-lyase is beneficial to look for a possible cause of the emergence of the flexible enantioselectivity [32]. Tyrosine phenol-lyase or TPase catalyzes the reversible hydrolytic cleavage of L- tyrosine or L- tryptophan to phenol or indole, respectively. They are very similar in sequence and structure. Although they are known to bind various amino acids, they show strict specificity for their respective physiological substrates. When enzymatic cleavage of the side chain is impossible, the bound amino acid acts as a reversible competitive inhibitor. The chemical transformation in the active site ends with the generation of the corresponding quinonoid intermediate in this case. If the same thing happens to D-Ser, the quinonoid intermediate will become a dead end.

D-Ser is a competitive inhibitor in the absence of TAP, even if D-Ser binding in the active site can be assumed to induce the conformational changes that depend on the structure of the side chain and affect the relative orientation of the C^{α}-H of the external aldimine and the ε-amino group of the acceptor lysine residue. When TAP is added, it produces small but enough conformational change to shift from the internal aldimine bond between the lysine residue and PLP to the external aldimine bond between D-Ser and PLP. The additional conformation change induces a change in their relative orientation that is certainly capable of promoting the deprotonation of C^{α}-H to produce α-aminoacrylate intermediate. As a result, D-Ser can work as active substrate to synthesize L-Trp. A superimposition of active-site residues taken from the spatial structures of tyrosine phenol-lyase in complex with 3-(4'-hydroxyphenyl)propionic acid and TPase holoenzyme was presented on the basis of crystallographic studies [33, 34]. According to this scheme, the C^{α}-H of D-Ser occupies the symmetrical mirror opposite position to the one of four L-Ser residues even though the other three residues bound to the α-carbon are superimposing one another. Since a steric configuration of D-Ser cannot superimpose one of L-Ser, this non-superimposition makes it difficult to explain why D-Ser becomes active. We previously reported D-His inhibited only D-Trp degradation but had no effect on L-Trp degradation [35]. This sparks an idea that the active site of D-Ser is partially independent of that of L-Ser, also making us imagine something like a pocket into which D-Ser slides in the immediate vicinity of L-Ser bound to PLP in the active site. Studies on the crystal structure of L- or D-serine dehydratase help to interpret our results [36, 37]. In the case of L-serine dehydratase, Lys41 abstracts the proton from the C^{α} of L-serine in a concerted fashion, whereas in D-serine dehydratase Thr168 is responsible for the abstraction. The role of the proton abstraction from the C^{α} of L-tryptophan is played by Lys270 in TPase. If TPase is the same as the serine dehydratase, the abstraction of that of D-Ser will be performed by any other possible amino acid with high basicity in the above-mentioned pocket. Thyrosine phenol-lyase has been demonstrated to have two distinct conformations that bind L-alanine and D-alanine, respectively [38]. Here we hypothesize a small conformational change in the enzyme from the native L-conformation to an alternative D-conformation at high TAP concentration. Since L-Ser binds to the L-conformation and adversely the D-conformation binds to the other. Since they lie next to each other, they inhibits competitively in the absence of TAP, but they move slightly away from each other in the presence of TAP. Consequently, the competitive inhibition attenuate to push D-Ser up to active substrate. This working hypothesis suggests the switch from L-conformation to D-conformation is controlled by TAP, offering interesting insights into the origin of homochirality. Today's dilute saline environment in aqueous solution may give an optimal condition for the exclusive use of the L-conformation, and consequently the L-amino acid may become dominant in contemporary biological world.

Conclusion

TPase produces pyruvate from D-Ser *via* a β-elimination reaction pathway in the presence of TAP, but when indole is present, it can synthesize L-Trp *via* β-replacement without pyruvate production. The reaction process of α-proton abstraction from D-Ser and indole association is quite complicated because of the involvement of the different acid-base catalyses of side chains of amino acids in TPase. D-Ser goes through the complex process of acid-base catalysis to be synthesized into L-Trp *via* several intermediates. It will be significant to identify their intermediates to understand this reaction mechanism in the presence of TAP. These intermediates will be detected and characterized by spectroscopic analysis in our future study. We break away from the conventional idea that enzyme enantioselectivity has both absolute stability and immutability. The enantioselectivity is underpinned by a delicate balance between conformation and saline environment. Although we have yet to understand how the exclusive selection of L-amino acid, that is, the origin of homochirality occurred in an era of early earth, our study suggests the saline environment is one significant factor to produce chiral homogeneity. Our results demonstrate TAP activates D-Ser as an active substrate in synthesis reaction, offering a good clue to solve the difficult problem on chiral homogeneity. Tryptophanase-catalyzed L-tryptophan synthesis system from D-serine is an attractive model for the study of the origin of homochirality.

Acknowledgments

The authors would like to thank the Project Research Program of the Kyoto University Research Reactor Institute, and JSPS KAKENHI Grant Number 24570247 for supporting partly our research.

REFERENCES

[1] Bonner, W. A. The origin and amplification of biomolecular chirality. *Orig. Life Evol. Biosph.*, 1991, 21, 59–111.

[2] Cohen, B. A., Chyba, C. F. Racemization of meteoritic amino acids. *Icarus*, 2000, 145, 272–281.

[3] Lazcano, A., Miller, S. L. How long did it take for life to begin and evolve to cyanobacteria? *J. Mol. Evol.*, 1994, 39, 546–554.

[4] Raulin-Cerceau, F., Maurel, M. C., Schneider, J. From panspermia to bioastronomy, the evolution of the hypothesis of universal life. *Orig. Life Evol. Biosph.*, 1998, 28, 597–612.

[5] Arrhenius, S. *Worlds in the Making: The Evolution of the Universe*, Harper, New York, 1908.

[6] Bada, J. L., Miller, S. L. Racemization and the origin of optically active organic compounds in living organisms. *Biosystems*, 1987, 20, 21–26.

[7] Aponte, J. C., Tarozo, R., Alexandre, M. R., Alexander, C. M. O., Charnley, S. B., Hallmann, C., Summons, R. E., Huang, Y. Chirality of meteoritic free and IOM-derived monocarboxylic acids and implications for prebiotic organic synthesis, *Geochim. Cosmochimi. Acta*, 2014, 131, 1-12.

[8] Lahav, M. Question 4: Basic questions about the origin of life: On chirobiogenesis. *Orig. Life Evol. Biosph.*, 2007, 37, 371–377.

[9] Pizzarello, S., Weber, A. L. Prebiotic amino acids as asymmetric catalysts. *Science*, 2004, 303, 1151, doi:10.1126/science.1093057.

[10] Bonner, W. A. Chirality and life. *Orig. Life Evol. Biosph.*, 1995, 25, 175–190.

[11] Reist, M., Carrupt, P. A., Francotte, E., Testa, B. Chiral inversion and hydrolysis of thalidomide: Mechanisms and catalysis by bases and serum albumin, and chiral stability of teratogenic metabolites. *Chem. Res. Toxicol.*, 1998, 11, 1521–1528.

[12] Fujii; N. D-amino acids in living higher organisms. *Orig. Life Evol. Biosph.*, 2002, 32, 103-127.

[13] Newton, W. A., Snell, E. E. Catalytic properties of tryptophanase, a multifunctional pyridoxal phosphate enzyme. *Proc. Natl. Acad. Sci. US*, 1964, 51, 382–389.

[14] Shimada, A., Nakamura, I. Degradation of D-tryptophan by tryptophanase under high salt concentration. *Viva Origino*, 1992, 20, 147–162.

[15] Shimada, A., Kogure, H., Shishido, H., Nakamura, I. Reaction pathway of tryptophanase degrading D-tryptophan, *Amino Acids*, 1997, 12, 379-383

[16] Shimada, A., Ozaki, H., Saito, T., Fujii, N. Tryptophanase-catalyzed L-tryptophan synthesis from D-serine in the presence of diammonium hydrogen phosphate. *Int. J. Mol. Sci.*, 2009, 10, 2578–2590.

[17] Kulikova, V. V., Zakomirdina, L. N., Bazhulina, N. P., Dementieva, I. S., Faleev, N. G., Gollnick, P. D., Demidkina, T. V. Role of arginine 226 in the mechanism of tryptophan indole-lyase from *Proteus vulgaris*. *Biochemistry*, 2003, 68, 1181–1188.

[18] Shimada, A. Activity on D-tryptophan attributable to slight conformational change of tryptophanase in highly concentrated ammonium phosphate solution. In: *Enzymes Involved in the Metabolism of D-Amino Acids: Practical Methods and Protocols*, Nova Science Publishers, New York, 2010, Volume 4, pp. 173–192.

[19] Isupov, M. N., Antson, A. A., Dodson, E. J., Dodson, G. G., Dementieva, I. S., Zakomirdina, L. N., Wilson, K. S., Dauter, Z., Lebedev, A. A., Harutyunyan, E. H. Crystal structure of tryptophanase, *J. Mol. Biol.*, 1998, 276, 603-623.

[20] Shimada, A., Ozaki, H., Saito, T., Fujii, N. Reaction pathway of tryptophanase-catalyzed L-tryptophan synthesis from D-serine. *J. Chromatogr. B*, 2011, 879, 3289–3295.

[21] Miles, E. W. Tryptophan synthase: a multienzyme complex with an intramolecular tunnel, *The Chemical Record*, 2001, 1, 140-151.

[22] Bentley, R. Diastereoisomerism, contact points, and chiral selectivity: A four-site saga. *Arch. Biochem. Biophys.*, 2003, 414, 1–12.

[23] Faleev, N. G., Ruvinov, S. B., Demidkina, T. V., Myagkikh, I. V., Gololobov, M. U., Bakhmutov, V. I., Belikov, V. M. Tyrosine phenol-lyase from *Citrobacter intermedius*, *Eur. J. Biochem.*, 1988, 177, 395-401.

[24] Phillips, R. S., Demidkina, T. V., Faleev, N. G. Structure and mechanism of tryptophan indole-lyase and tyrosine phenol-lyase. *Biochim. Biophys. Acta*, 2003, 1647, 167–172.

[25] Shimada, A., Fujii, N., Saito, T. Tryptophanase activity on D-tryptophan. In: Paliyi, G., Zucchi, C., Caglioti, L. (Eds.), *Progress in Biological Chirality*, Elsevier (Oxford), 2004, pp. 321-327.

[26] Kortemme, T., Ramirez-Alvarado, M., Serrano, L. Design of a 20-Amino Acid, Three-Stranded β-Sheet Protein, *Science*, 1998, 281, 253-256.

[27] Kuroda, Y., Maeda, Y., Sawa, S., Shibata, K., Miyamoto, K., Nakagawa, T. J. Effects of detergents on the secondary structures of prion protein peptides as studied by CD spectroscopy, *J. Pept. Sci.* 2003, 9, 212-220.

[28] Shimada, A., Shishido, H., Kogure, H., Nakamura, I., Fuji, N., Akaboshi, M. Tryptophanase-catalyzed D-tryptophan degradation reaction and its significance for chiral homogeneity, In: M. Akaboshi (Eds.), *The Role of Radiation in the Origin and Evolution of Life*, Kyoto University Press, Kyoto, 2000, pp. 243-257.

[29] Ku, S., Yip, P., Howell, P. L. *Acta Crystallogr. D Biol. Crystallogr.* Structure of *Escherichia coli* tryptophanase, 2006, D62, 814-823.

[30] Kogan, A., Gdalevsky, G. Y., Cohen-Luria, R., Goldgur, Y., Phillips, R. S., Parola, A. H., Almog, O. Conformational changes and loose packing promote *E. coli* Tryptophanase cold lability, *BMC Struct. Biol.* 2009, 9, 65-76.

[31] Shimada, A. Role of ammonium phosphates in tryptophanase activity toward D-tryptophan, In: R. Konno, H. Brückner, A. D'Aniello et al. (Eds.), *D-Amino Acids: A New Frontier in Amino Acid and Protein Research - Practical Methods and Protocols*, Nova Science Publishers, New York, 2008, pp. 591-607.

[32] Demidkina, T. V., Antson, A. A., Faleev, N. G., Phillips, R. S., Zakomirdina, L. N. Spatial structure and mechanism of tyrosine phenol-lyase and tryptophan indole-lyase, *Mol. Biol.* (Moskow), 2009, 43, 295–308.

[33] Sundararaju, B., Antson, A. A., Phillips, R. S., Demidkina, T. V., Barbolina, M. V., Gollnick, P., Dodson, G. G., Wilson, K. S. The crystal structure of *Citrobacter freundii* tyrosine phenol-lyase complexed with 3-(4'-hydroxyphenyl)propionic acid, together with site-directed mutagenesis and kinetic analysis, demonstrates that arginine 381 is required for substrate specificity. *Biochemistry*, 1997, 36, 6502–6510.

[34] Phillips, R. S., Demidkina, T. V., Faleev, N. G. Structure and mechanism of tryptophan indole-lyase and tyrosine phenol-lyase. *Biochim. Biophys. Acta*, 2003, 1647, 167–172.

[35] Shimada, A. Flexible enantioselectivity of tryptophanase attributable to benzene ring in heterocyclic moiety of D-tryptophan, *Life*, 2012, 2, 215-228.

[36] Yamada, T., Komoto, J., Takata, Y., Ogawa, H., Pitot, H C., Takusagawa, F. Crystal structure of serine dehydratase from rat liver. *Biochemistry*, 2003, 42, 12854–12865.

[37] Urusova, D. V., Isupov, M. N., Antonyuk, S., Kachalova, G. S., Obmolova, G., Vagin, A. A., Lebedev, A. A., Burenkov, G. P., Dauter, Z., Bartunik, H. D., Lamzin, V. S., Melik-Adamyan, W. R., Mueller, T. D., Schnackerz, K. D. Crystal structure of D-serine dehydratase from *Escherichia coli*, *Biochim. Biophys. Acta*, 2012, 1824, 422–432.

[38] Chen, H., Phillips, R. S. Binding of phenol and analogues to alanine complexes of tyrosine phenol-lyase from *Citrobacter freundii*: Implications for the mechanisms of α, β-elimination and alanine racemization. *Biochemistry* 1993, 32, 11591–11599.

In: New Developments in Tryptophan Research ISBN: 978-1-63483-025-6
Editor: Victoria Hayes

Chapter 5

TRYPTOPHAN MOLECULE FRAGMENTATION IN COLLISIONS WITH LOW-ENERGY ELECTRONS

***A. V. Snegursky*[*,1], *J. Tamuliene*[2], *L. G. Romanova*[1], *V. S. Vukstich*[1] and *A. V. Papp*[1]**

[1]Institute of Electron Physics, Ukrainian National Academy of Sciences, Uzhgorod, Ukraine

[2]Vilnius University, Institute of Theoretical Physics and Astronomy, Vilnius, Lithuania

ABSTRACT

The present chapter is devoted to the complex experimental and theoretical studies of the tryptophan ($C_{11}H_{12}N_2O_2$) molecule fragmentation in its collisions with low (<200 eV) and high (9.5 MeV) energy electrons. The mass-spectrometric technique was used to identify the ionic fragments produced as well as to trace their yield dynamics allowing the appearance energies of the main ionic fragments to be found. The experimental apparatus using a magnetic mass-analyzer has been modified and updated according to the purpose of the present studies.

A special attention has been paid to the energy and geometric characteristics of the fragments produced. A number of doubly-charged fragment ions (e.g., $C_8H_5N^{2+}$, $C_9H_7N^{2+}$ and $C_9H_9N^{2+}$) have been studied

[*] E-mail: snegursky.alex@gmail.com

for the first time. The main emphasis has been given to the problem of evaluating the possible mechanisms of the initial molecule fragmentation using the generalized gradient approximation for the exchange-correlation potential in the density functional theory (DFT). The structures of the parent molecule and its relevant fragments have been optimized globally with no symmetry constraint. The bond orders and bond lengths of the tryptophan molecule have been investigated to find the weaker bonds to be destructed. In order to model the fragmentation processes, the possible fragment anions, cations and fragments with a zero charge both with and without geometry optimization have been evaluated. The dissociation energies have been calculated as the difference between the total energy of the tryptophan molecule and the sum of the energies of the fragments predicted. The geometrical parameters of the initial tryptophan molecule rearrangement have also been analyzed.

It has been confirmed that the main channel of the tryptophan molecule dissociation is related to the simple cleavage of the C_α–C_β bond of the initial molecule. This bond dissociation is the most probable when electron energies slightly exceed the tryptophan molecule ionization potential. When the electron energy is sufficient to ionize deeper molecular orbitals, first of all, the lone pair ones of nitrogen and oxygen, dissociation of the O and N atom neighboring bonds takes place, and, as a consequence, more fragmentation channels occur. However, the number of such dissociation channels is not significant because the aromatic group stabilizes the positive charge and reduces the fragmentation. The results obtained are being compared to those found in the relevant databases and literature.

Introduction

An electron-molecule interaction is a fundamental and very important process involved in various fields, e.g., in radiation biology. In this area, low-energy electrons represent the most predominant species formed during a very short time after the high-energy ionizing quanta deposition into a biological medium [1]. Once produced, they are able to destruct the environing biological molecules (say, DNA and proteins) inducing chromosome aberrations such as cancer mutations, genetic transformations, etc. [2, 3]. Thus, it is worth to investigate in detail the underlying mechanisms of the above processes by studying the degradation of the biosystem sub-units under the low-energy electron impact.

In our investigation, we studied interactions of electrons with a tryptophan molecule belonging to essential amino acids in order to probe the intrinsic properties of the molecule and trace its change(s) under the electron impact. Tryptophan is the metabolic precursor of serotonin, which plays an important role in a sleeping process, and is necessary for normal growth in infants and for nitrogen balance in adults. In the standard genetic code, it is encoded as the codon UGG.

This molecule, like other amino acids, has several conformers. The L-stereoisomer of tryptophan is used in structural or enzyme proteins, while the D-stereoisomer is occasionally found in naturally produced peptides (for example, in the marine venom peptide contryphan) [4]. The D-isomer of the molecule is not utilized by humans [5], while L-tryptophan is used for coping with insomnia, sleep apnea, depression, anxiety, facial pain, a severe form of a premenstrual syndrome called the premenstrual dysphoric disorder (PMDD), a smoking cessation, grinding teeth while sleeping (bruxism), attention-deficit-hyperactivity disorder, Tourette's syndrome, as well as for improving the athletic performance [6].

Additionally, L-tryptophan is naturally found in animal and plant proteins. However, the studies of the interaction of low-energy particles with tryptophan are scarce: there are some data on the molecule fragmentation when the negatively charged species are observed (see, e.g., [1]) or on the fragmentation of the radical-cationic tryptophan [7].

To better understand the initial molecular processes taking place during the interaction of radiation with a living system, the results of studying the process of the positively charged fragment formation under electron impact using the mass-spectrometric technique are presented here. Tryptophan is unique among amino acids since its side chain contains an aromatic indole condensed ring, which has the lowest ionization energy among all the amino acid side chains [1].

A theoretical method was used to predict both possible fragmentation channels and processes that could occur under electron impact, involving the additional ones, such as dehydration and the H atom migration.

The theoretical results were compared with the experimentally measured data aiming to check the theoretical predictions made, choose the dominant fragmentation pathways, and describe fragmentation processes more correctly.

Hence, the goal of our studies was to elucidate the major channels of the tryptophan molecule fragment formation (fragmentation) from one of the most stable conformer due to electron impact.

EXPERIMENTAL

In the present research, a magnetic mass-spectrometer MI-1201 was applied to study the production of a number of fragments resulted from the tryptophan molecule dissociative ionization by electron impact. The above apparatus operates within the *m/z*=1–600 a.m.u. mass-to-charge range providing high sensitivity (~10^{-16} A) and resolution (±0.25 a.m.u.), ensuring, thus, reliable separation and detection of ionic fragments with different molecular masses *m*. The detailed description of the experimental apparatus and technique used in this research can be found in our previous papers (see, e.g., [8, 9]).

A schematic diagram of the experimental mass-spectrometer-based apparatus used in the present study is shown in Figure 1.

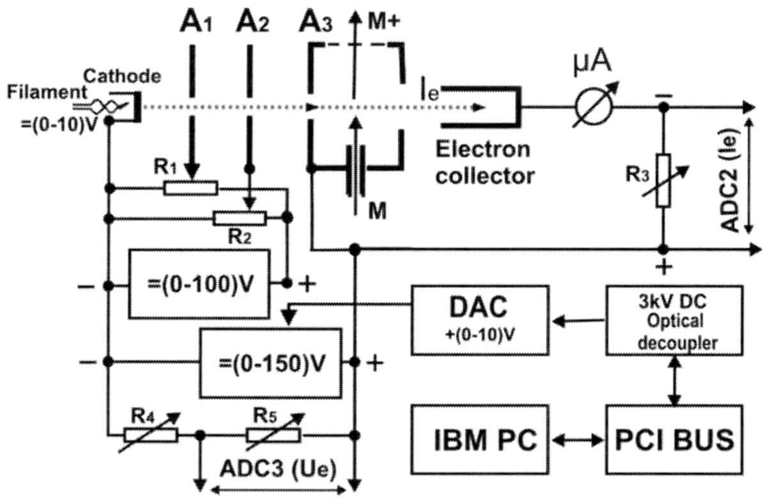

Figure 1. Schematic diagram of the experimental apparatus.

The beam of the parent tryptophan molecules *M* (see Figure 1) was generated using a modified resistive-oven based effusion source providing the target molecule concentrations in the ion source of the mass-spectrometer of about 10^{10} molecule/cm^3. This beam entered the ionization chamber normally to the electron beam. The working temperature of the molecular beam source being controlled by a K-type thermocouple with a possibility to apply the platinum resistance temperature sensor Pt100 was varied up to 170°C. This allowed us to determine the temperature dependences of the fragment yields in order to avoid thermal destruction (fragmentation) of the parent molecule in the effusion source. A modified three-electrode electron gun provided an ionizing electron current of 30–50 μA within the incident electron energy range from the threshold up to 150 eV. The ions produced in the ion source were extracted by the electric field, entered the magnetic ion separator and were detected by an electrometer interfaced with the signal processing system developed by us (see below). The electron energy scale was calibrated with respect to known ionization thresholds for the argon atom and nitrogen molecule ensuring the energy calibration accuracy of $\pm$0.1 eV.

The experimental procedure was as follows. First, both electron and molecular beam sources were put into operation and, after reaching the optimal conditions of beam generation, the mass spectrum of the molecules under study was measured at the 66 eV electron energy. Since amino acid molecules may undergo strong thermal decomposition, the above mass spectra were measured serially at different molecular beam source oven temperatures. The temperature dependence of the yield of the ionized fragments with the mass m=77 a.m.u. (this mass corresponds to the $C_6H_5^+$ and $C_5H_3N^+$ isobaric ions) produced of the parent tryptophan molecule is shown in Figure 2. Note, that the same dependences for other most prominent tryptophan molecule ionic fragments have the same form.

This dependence, being plotted on the semi-logarithmic scale, demonstrates a distinct linear behavior. It may be concluded from this dependence that the molecular beam source temperature up to 170°C ensures reliable detection of the ion fragment yield with no thermal degradation of the initial molecule. In other words, the contribution of the parent molecule thermal degradation to the ion fragment(s) yield could be undoubtedly neglected in our experiment.

The masses of the fragments produced were then fixed and the energy dependences of the relevant ion yields were measured. The energy was varied with an energy step of 0.1–0.3 eV what enabled the threshold areas of the dissociative ionization functions for all the fragments under study to be measured. In this case the problem of the incident electron energy scale calibration is, certainly, very significant. In order to calibrate the above scale we have measured firstly the threshold areas of the ionization functions for the two test gases – Ar and N_2. The experimental ionization thresholds were determined by means of a fitting technique suggested by T. Maerk's group (see, e.g., our earlier paper [10]) and used in our previous studies. This technique is based on the least-square method approximation using the Marquardt-Levenberg algorithm. The absolute value of the threshold of the Ar atom ionization energy obtained by us (E=15.76 eV) agrees well with that known from other studies and reported in a number of databases (see, e.g., [11]). The same is valid for the relevant data on the nitrogen molecule. The ionization threshold value for this molecule, determined by us, i.e., 15.56 eV, agrees extremely well with the NIST data [11]. Thus, our technique obviously ensured reliable measurements of the absolute energy thresholds for the processes under study.

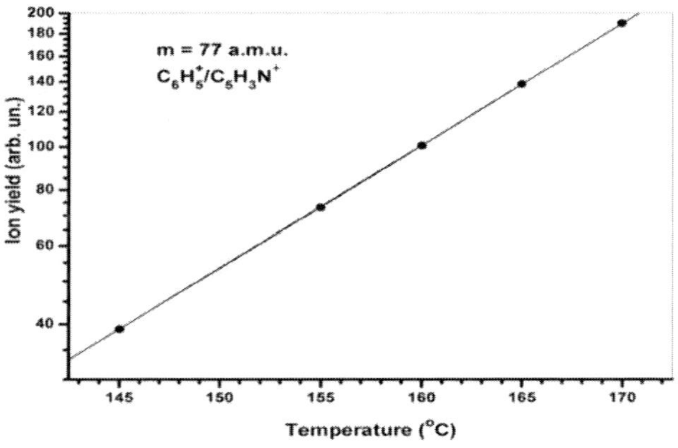

Figure 2. Temperature dependence of the $C_6H_5^+$ and $C_5H_3N^+$ fragments yield (*m*=77 a.m.u.).

In order to carry out deeper studies of the ionizing radiation influence on the tryptophan molecule fragmentation, we have measured the mass-spectrum of the initial molecule after its interaction with the high-energy (9.5 MeV) electron beam. This beam was generated by the electron accelerator (an M-30 microtron). The sample in the form of a fine tryptophan crystalline powder was irradiated by the electron beam up to the dose of 5 kGy (an integral flux being $1.8 \cdot 10^{13}$ e/cm^2 at the average fluence intensity of $1.6 \cdot 10^{10}$ e/cm$^2 \cdot$s).

In this experiment, the presence of the Bremsstrahlung radiation is unavoidable and contributes to the irradiation results, too. In order to check the absence of thermal degradation of the sample under the action of an intense electron beam, the sample temperature during exposure was monitored by a copper-constantan thermocouple, which showed that the beam-induced heating did not exceed 1°C. An irradiation field of the 5×8 mm size was cut from the electron beam cross-section at the accelerator exit with the help of a tungsten target and collimator. The inhomogeneity of the radiation field close to the sample did not exceed 1%. Note that the average fluence intensity differs significantly from that of the peak firstly due to the pulsed character of the electron beam produced by the microtron with the 2,000 duty ratio, and, secondly, due to the formation of the so-called beam bunches related to the narrow phase region of the stable acceleration.

A Multi-Purpose Detection System

It is a well-known fact that the magnetic mass-spectrometers MI1201 received a wide application in different scientific centers as early as in the 70-s of the previous century [8]. Certainly, nowadays some units of such quite expensive research complexes have exhausted their initial resource. In order to improve the apparatus operation, i.e., to realize the system of automating the mass-spectrometric experiment by controlling the mass-spectrum scanning and ion source operation, as well as other parameters (i.e., peak intensity, mass value, working substance temperature) measuring, we have earlier modified the MI1201 mass spectrometer apparatus (see, e.g., [8, 9]). The detection system was built on the IBM PC basis with the use of the 12-bit digit-to-analog converters (DACs), analog-to-digit converters (ADCs) and a specially developed mass spectrum scanning system [12, 13].

In view of a rapid development of the electronic element basis, we decided to carry out the next modernization of the mass-spectrometer controlling and detecting system. The new version of this system reflects a principally novel approach not related to a particular PC and using a multi-processor modular system as the main control device. Each functional module is a separate microcontroller-based unit that controls the operation of certain analog circuits and is interfaced to a common digital network. Such a solution is frequently used in modern electronics. In this system, a PC is connected via an USB interface and uses only the digital signals processed previously by relevant functional modules.

The digital network operates in accordance with the Controller Area Network (CAN) protocol. The CAN is a serial bus microcontroller network that connects devices, sensors and actuators in a system or sub-system for real-time control applications. There is no addressing scheme used in the controller area networks as it is in the case of a conventional addressing in other networks (such as the Ethernet). The messages are broadcasted to all the nodes in the network using an identifier unique to the entire network. Based on the identifier, the individual nodes decide whether or not to process the message and determine the priority of the message in terms of competition for bus access. This method allows the uninterrupted transmission to be realized if any conflict is being detected, while the Ethernet stops the transmission upon a conflict occurrence. All the functional modules have an optical decoupling through the network with maximally allowable voltage difference of 5 kV. Figure 3 shows the general flow-chart of the controlling and detecting system using the CAN network.

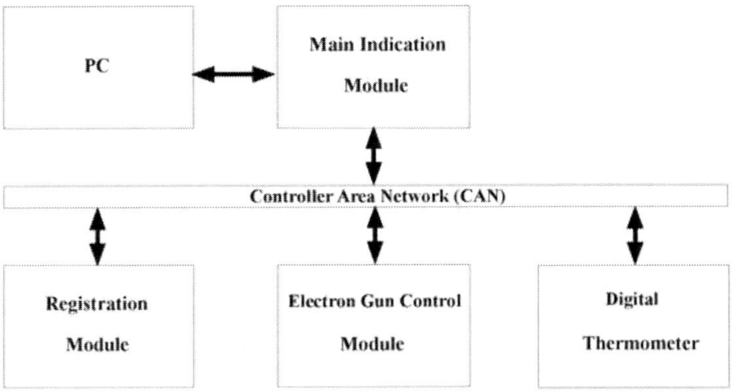

Figure 3. General flow-chart of the controlling and detecting system.

An indication module is a principal unit that reflects all the main parameters of the system using the Winstar Alphanumeric LCD display capable of displaying up to 160 characters and providing matching between the CAN network and the USB/RS232 interface. It is based on a high-performance low-power Atmel 8-bit AVR RISC-based microcontroller and combines the 128 KB ISP flash memory with the read-while-write capabilities, 2 KB EEPROM, 4 KB SRAM, 53 general purpose I/O lines, 32 general-purpose registers, and the CAN controller (V2.0A/V2.0B compliant). The USB interface operates via the FTD232 bus driver. The availability of a keyboard allows the parameters of other modules to be modified without the need of using a PC. The function block diagram of the indication module is shown in Figure 4.

As mentioned above, the MI1201 mass-spectrometer includes the electron-impact ion source with a three-electrode electron gun used to produce low-energy electrons. The main task of this module is to provide high stability of electric potentials at the ion source electrodes and to ensure electron emission current stabilization.

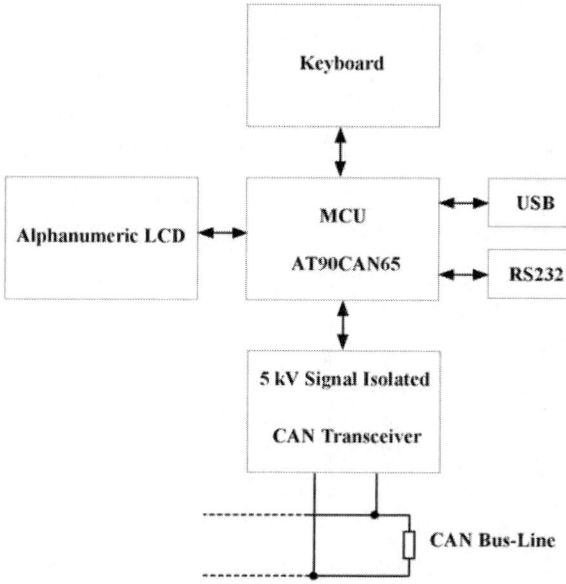

Figure 4. Function block diagram of the indication module.

The electron gun control module has three highly-stable DC voltage power supplies based on the high precision, zero-drift, rail-to-rail operational amplifiers. The values of the voltages are set by the precise 16-bit DACs and are controlled by the 16-bit sigma-delta (ΣΔ) ADCs. Both ADCs and DACs) communicate with a microcontroller via a Serial Peripheral Interface (SPI) as an interface bus commonly used to send data between microcontrollers and small peripherals such as shift registers, sensors, and the SD cards. All the analog inputs have an overvoltage protection that excludes the input circuit damage. This module uses an ADR4540 low-noise voltage reference featuring ±0.02% initial error with the excellent temperature stability. All the voltage power supplies are controlled by an ATMEL microcontroller. A special advantage of this solution is the electron emission current stabilization and variation. Figure 5 illustrates the function block diagram of the electron gun control module.

The detection and scanning module developed in this work is absolutely new as compared to the conventional ones. In the earlier version, the digital scanning generator was used with a high-power amplifier not free from a number of shortcomings, one of them being the lack of the output current feedback not allowing high magnetic inductance stability to be achieved for a long time. A voltage-to-frequency converter was used to measure the ion current. Such converter is quite outdated and has worse parameters as compared to modern ΣΔ ADC. The newly designed module has no thermostat or other elements characterized by a strong temperature dependence of their parameters. This enables the mass-spectrometer operation to be organized with no heating up of electronic elements.

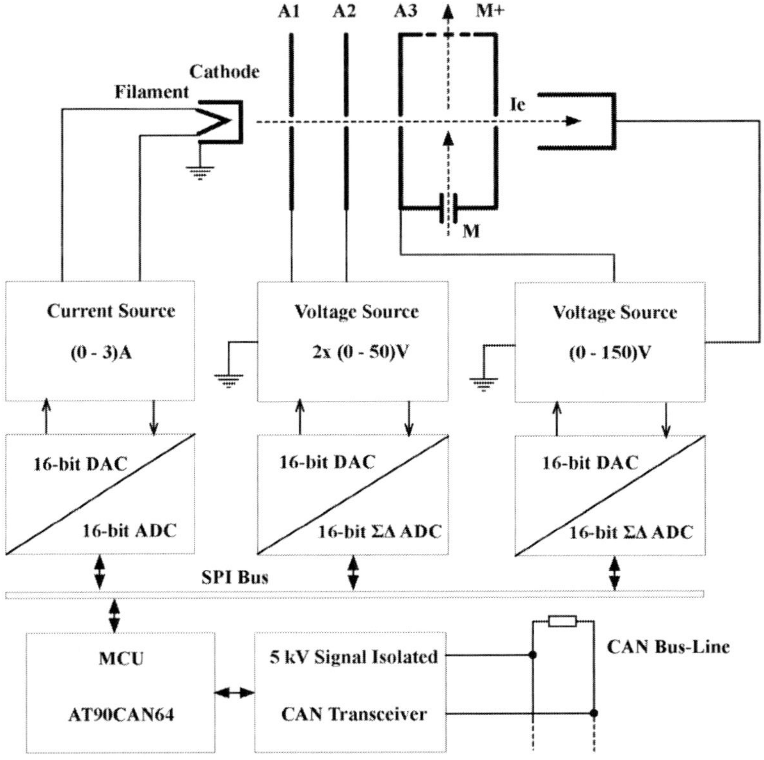

Figure 5. Function block diagram of the electron gun control module.

To measure the ion current preamplified by an electrometric amplifier, the detection module uses a converter with the maximally allowable resolution, i.e., the 24-bit ΣΔ ADC ADS1222. To simplify the system architecture in general, the detection module comprises a high-power current power supply for the mass analyzer electromagnet and a precise 18-bit DAC MAX5318 with the ultra linear characteristics that provide the mass-spectrum scanning. A high-power current supply is based on the Maxim Integrated ultra-precision, low-noise, low-drift amplifiers that offer near zero DC offset and drift by using the autocorrelating zeroing techniques. The output current is varied from 0 to 10 A with a high accuracy.

The function block diagram of the detection and scanning control module is shown in Figure 6.

A combined thermometer is the next element of this system. As mentioned above, the molecular beam is produced by an effusion source that requires permanent temperature control to provide stable molecule concentration. Precise measuring of the working medium temperature is also necessary when studying the temperature dependences of ion yields. The thermometer is absolutely flexible and allows the temperature to be measured using both the K-type thermocouple and the platinum resistance temperature sensor Pt100.

Similarly to the above devices, it is based on the AMTEL microcontroller and has a Winstar Alphanumeric LCD display capable of indicating up to 24 characters. When using the thermocouple, the temperature is measured by a cold-junction compensated thermocouple digital converter MAX31855. The data output format is a 14-bit, SPI-compatible read-only one. Its temperature resolution is ±0.25°C within the temperature range from –270°C to +1,800°C with the ±2°C accuracy in the temperature range from –200°C to +700°C for the K-type thermocouple.

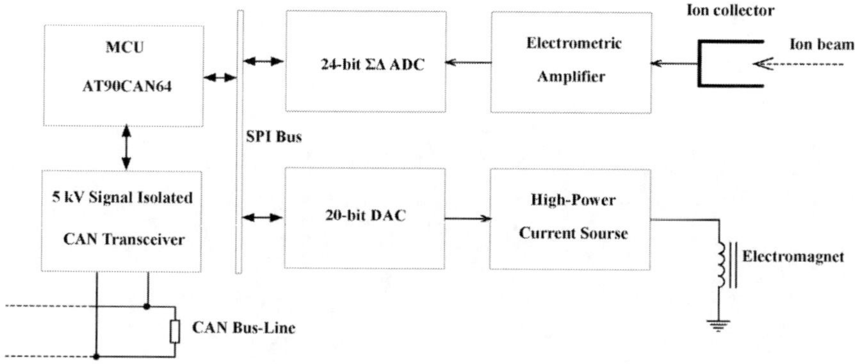

Figure 6. Function block diagram of the detection and scanning control module.

In case of a platinum resistor, a MAX31865 RTD (Resistance Temperature Detectors) digital converter is used as being compatible to the 2-, 3-, and 4-wire sensor connections. It is based on a 15-bit ADC with the nominal temperature resolution of ±0.03125°C. The total accuracy for all operating conditions is ±0.5°C (0.05% of a full scale). Figure 7 shows the structural diagram of the digital thermometer.

Theoretical

The structures of the molecule and its fragments have been studied by the DFT theory. We have used the Becke's three-parameter hybrid functional applying the non-local correlation provided by Lee, Yang, and Parr (B3LYP) [14, 15] with the cc-pVTZ basis [16] as the method allowing accurate results to be achieved with the effective use of the computational resources. The structures of the molecule isomers and their fragments under study have been globally optimized. The vibration spectrum has also been evaluated to ensure that the equilibrium geometry is obtained and to predict the possible elongation of bonds and the change of the angle with the aim to make the decision on the most probable fragments production by electron impact. The zero-point energy has been taken into account in case of its sufficient influence.

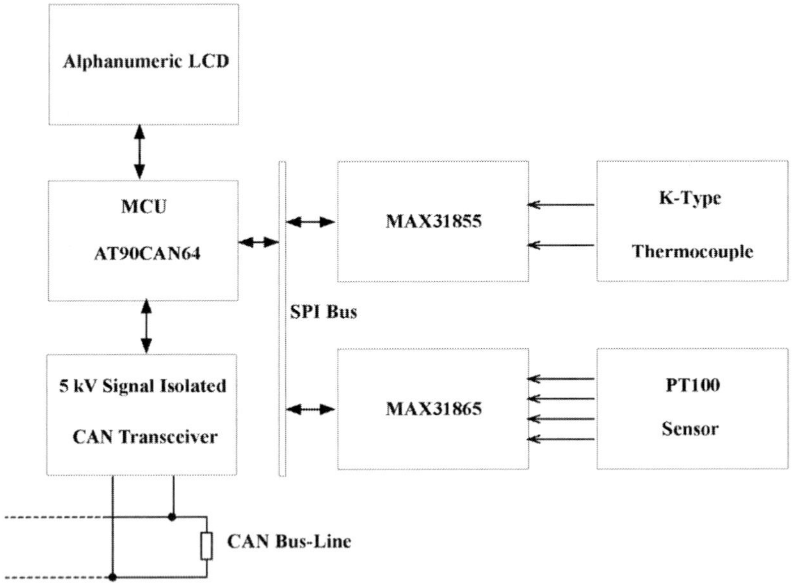

Figure 7. Structural diagram of the digital thermometer.

Along with the vibration spectrum analysis, the bond order and the bond length of the neutral and ionized trypthophan molecule have been investigated. The results of this analysis were used to find the weaker bonds to be destructed. In order to model the fragmentation processes, the production of different fragment anions, cations and fragments with a zero charge has been analyzed to predict the influence of dissociation energy on the fragmentation processes. The appearance energies have been calculated as

$$\Delta E = \left| E_{mol} - \sum E_i \right|,$$

where E_{mol} is the total energy of the trypthophan molecule, while E_i are those of the fragments produced.

Aiming to evaluate the above influence, the structure of the tryptophan molecule fragments has been optimized, i.e., the fragments were allowed to reach their equilibrium geometry and the obtained energy (the lowest energy of the fragments) has been used to calculate that of the dissociation. While applying this methodology, we have taken into account the processes when the molecular ions formed with the energies just in excess of the ionization potential may not have the energy sufficient to be decomposed along the lowest energy pathway and will, thus, be collected and detected [16]. The Gaussian program packages [17] have been applied here.

RESULTS AND DISCUSSION

The tryptophan molecule conformer image and the atom numbers used in this chapter are presented in Figure 8. The main peculiarity of the tryptophan molecule as compared to other amino acid molecules is the presence of an aromatic indole moiety in the side chain. Indole is the π-excess 10 π-electron aromatic system that includes the lone pair electrons of the nitrogen atom increasing the electron density in the carbon atoms present in both the pyrrole and benzene parts of the molecule under study [18].

In general, the conformational picture of amino acids, in both the gas phase and isolated in low-temperature matrixes, is considerably complicated. The amino acid molecules have low symmetry and a plenty of torsion vibration coordinates, due to which a large number of various low-energy conformers do exist. In the tryptophan conformers, different mutual orientations of the functional groups, the aliphatic chain and the indole fragment are inherent. The conformational analysis performed in [19, 20] resulted in finding 40 different forms of the tryptophan molecules as being the minima at the potential energy surface. In this case the total energy values for the 17 conformers lie in the range of 0.4 eV with respect to the energy of the most stable form. We have analyzed several structural isomers of the tryptophan molecule with the minimal total energy values (see Figure 9).

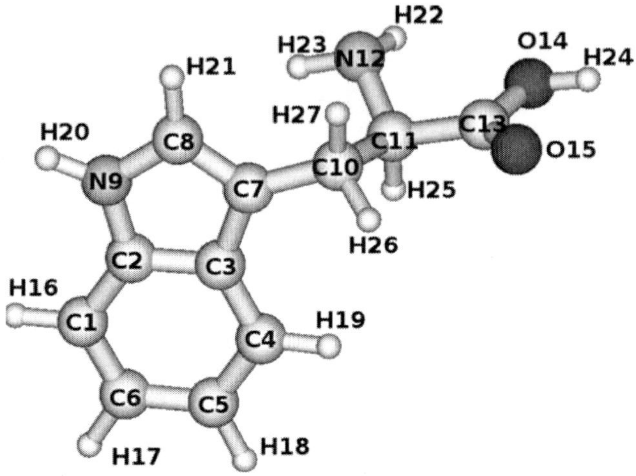

Figure 8. The tryptophan molecule image and the atom numbers used.

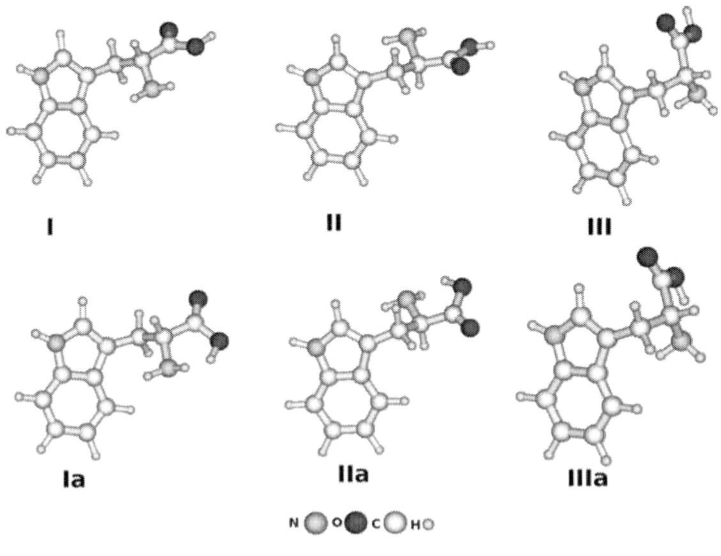

Figure 9. Tryptophan molecule conformers.

The main difference between the I and Ia, or II and IIa, or III and IIIa conformer structures is the position of the H atom of the OH group that leads to the most favorable position for the hydrogen bond formation.

The calculated hydrogen bond ($OH\cdots NH_2$) length corresponds to the values given in [21]. The above difference of the position of the H atoms leads to a higher stability of the conformers Ia, IIa and IIIa than that of I, II and III. Definitely, the total energy of the conformers, where H position is the same with respect of the nearest N atoms, lies within the range of 0.001 a.u (0.027 eV) (see Table 1).

On the other hand, the higher ionization energy values for the case of the conformers Ia, IIa and IIIa are, in our opinion, related to the presence of the intramolecular hydrogen bond that, probably, increases the electron density in the chiral carbon atom region. The positive induction effect, in turn, causes a positive mesomeric effect in the system of conjugated indole ring bonds resulting in the electron density increase and, as a consequence, in the increase of the molecule ionization energy.

Here we present the calculated appearance energies for the conformer II because the molar fraction of this conformer is the largest among the group of conformers without the intramolecular hydrogen bond.

In general, the incident free electron ionizes the target molecule to form a transitory ion. The parent precursor ion having internal energies below the dissociation energy remains stable [22]. In case when the internal energies are higher than the dissociation energy, the fragmentation processes start. The tryptophan molecule, apart of the indole ring with aromatic electron cloud, comprises two isolated functional groups involving heteroatoms with lone pair electrons, while the carbonyl group comprises the π-electrons. The ionization of this molecule with minimal energy consumption could proceed due to the electron loss from the indole π-electron ring, the non-bonding heteroatom n-orbitals and the elimination of the double-bond π- electron.

The ionization tends to cause the weakening of bonding within the ion as compared to the neutral precursor. A weaker bonding means longer bond lengths in the average indicating a higher tendency towards the bond dissociation. To estimate the changes in the tryptophan molecule geometry as the result of ionization, we have calculated both the bond lengths and the bond orders of the neutral and ionized tryptophan molecules after the geometry optimization. Implementing the Mulliken population analysis data [23], the weakest bonds in the tryptophan molecule were determined. The calculated bond lengths and the bond orders of both the neutral and ionized tryptophan molecules are listed in Table 2.

Table 1. Total energy (a.u.), ionization energy (eV) and molar fraction (%) of the tryptophan conformers at 433K

Conformer	I	II	III	Ia	IIa	IIIa
Total energy	-686.617	-686.618	-686.616	-686.621	-686.622	-686.623
Ionization energy	7.12	7.10	6.96	7.28	7.31	7.31
Mole part	3.8	11.9	1.7	14.4	31.8	36.6

Table 2. Bond lengths and bond orders before and after the tryptophan molecule ionization. Bond lengths for the indole moiety are underlined

Bond	Bond length, Å, Neutral	Bond length, Å, Ionized	Bond order, Neutral	Bond order, Ionized	Bond length difference, Å	Bond order difference
<u>C1-C2</u>	1.394	1.371	1.358	1.485	0.023	-0.127
<u>C1-C6</u>	1.384	1.409	1.536	1.365	-0.025	0.171
<u>C1-H16</u>	1.082	1.081	0.901	0.879	0.001	0.022
<u>C2-C3</u>	1.418	1.415	1.408	1.352	0.003	0.056
<u>C2-N9</u>	1.375	1.405	1.015	0.923	-0.030	0.092
<u>C3-C4</u>	1.402	1.405	1.358	1.340	-0.003	0.018
<u>C3-C7</u>	1.441	1.421	1.265	1.322	0.020	-0.057
<u>C4-C5</u>	1.082	1.391	1.53	1.446	-0.309	0.084
<u>C4-H19</u>	1.090	1.081	0.893	0.877	0.009	0.016
<u>C5-C6</u>	1.405	1.393	1.393	1.480	0.012	-0.087

Bond	Bond length, Å, Neutral	Bond length, Å, Ionized	Bond order, Neutral	Bond order, Ionized	Bond length difference, Å	Bond order difference
C5-H18	1.077	1.076	0.895	0.844	0.001	0.051
C6-H17	1.082	1.081	0.91	0.889	0.001	0.021
C7-C8	1.405	1.393	1.393	1.480	0.012	-0.087
C7-C10	1.498	1.485	0.911	0.931	0.013	-0.02
C8-N9	1.380	1.33	1.024	1.271	0.050	-0.247
C8-H21	1.090	1.081	0.893	0.877	0.009	0.016
N9-H20	1.000	1.009	0.837	0.806	-0.009	0.031
C10-C11	1.537	1.530	0.900	0.929	0.007	-0.029
C10-H26	1.090	1.091	0.934	0.928	-0.001	0.006
C10-H27	1.090	1.097	0.955	0.877	-0.007	0.078
C11-N12	1.464	1.464	0.923	0.999	0.000	-0.076
C11-C13	1.521	1.535	0.810	0.826	-0.014	-0.016
C11-H25	1.102	1.092	0.924	0.911	0.010	0.013
N12-H22	1.012	1.015	0.864	0.857	-0.003	0.007
N12-H23	1.013	1.013	0.869	0.860	0.000	0.009
C13-O14	1.203	1.202	1.987	1.930	0.001	0.057
C13-O15	1.356	1.343	1.018	1.029	0.013	-0.011
O14-H24	0.968	0.970	0.805	0.790	-0.002	0.015

Low-Energy Electron Impact Fragmentation

The tryptophan molecule mass-spectrum measured at the 66 eV electron energy (see Figure 10) is generally close to that quoted in the NIST database [11]. This spectrum is characterized by a high selectivity, since the intensities of the most distinct peaks related to the ions with the *m*=28, 77, 103 and 204 a.m.u masses do not exceed 7% of that of the peak with the *m*=130 a.m.u. mass that corresponds to the main channel of the molecular ion dissociation. It should be noted here that, under electron impact, a lot of fragments are produced with the mass-spectrum peak intensity not exceeding 3%. In general, this mass-spectrum is similar to those measured earlier for the tryptophan molecule in case of photoionization [25, 26] and electron-impact ionization [11]. However, one should note the relatively low-intensity peaks corresponding to the fractional masses *m*=57.5, 64.5, 65.5, 79.5 a.m.u (the latter being the most intense of them) and a diffuse peak at about *m*=81.5 a.m.u (all marked in dark tint) indicating the presence of the doubly charged ionic fragments and the ion decaying on its path from the ion source to the detector. We believe that these ions have been found by us experimentally for the first time.

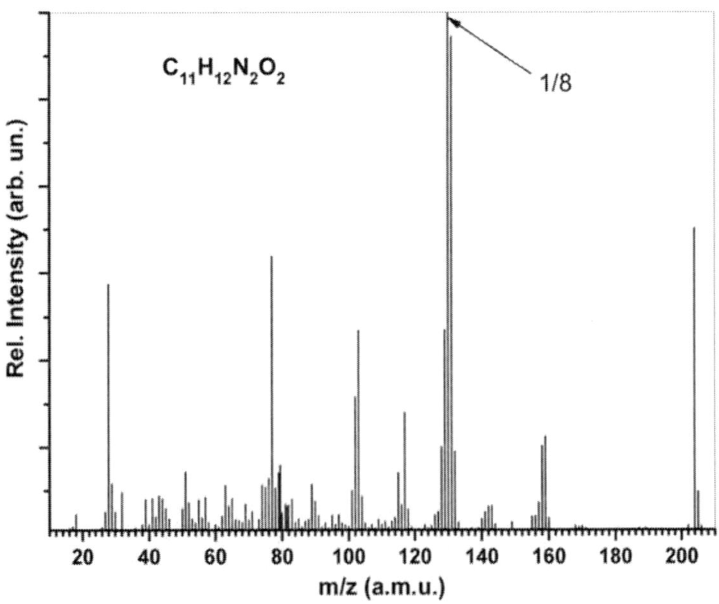

Figure 10. Tryptophan molecule mass-spectrum.

Obviously, the intensities of the aliphatic amine acid molecular ion peaks in the mass-spectra are very low due to the high probability of the decay processes initiated by the cation radical center localized at the amine group. However, due to the aromatic character of the side chain, tryptophan displays rather different pattern, and the intensity of the parent molecule signal in the mass-spectrum makes the determination of its ionization potential possible.

Using the above algorithm and the experimental initial molecule ionization curve, we have determined the ionization threshold E_i for the parent $C_{11}H_{12}N_2O_2$ molecule (*m*=204 a.m.u., see Figure 11a) and the appearance potential E_{ap} for the $C_9H_8N^+$ ion (*m*=130 a.m.u., Figure 11b). The value E_i=8.3±0.1 eV is slightly lower than that obtained in our previous studies for the molecules of some other aliphatic amino acids [27–30] and agrees well with that measured in the photoelectron spectroscopy experiment, i.e., 7.9 (π_1), 8.3 (π_2) and 9.8 eV for the first molecular orbitals (MO) of tryptophan [31].

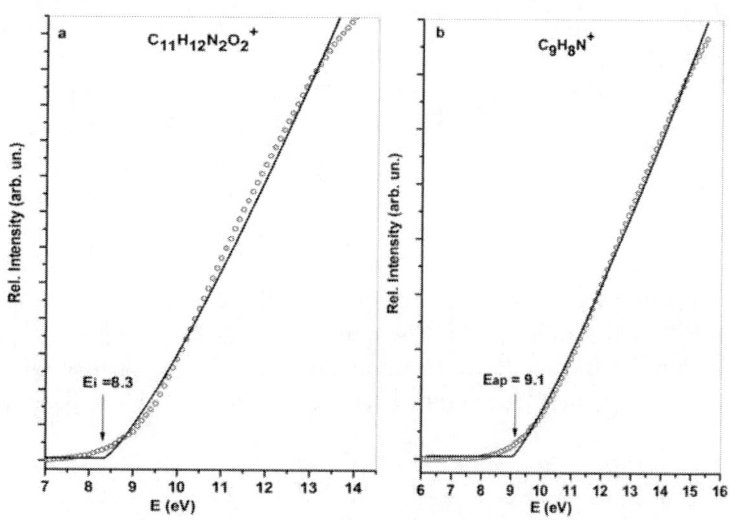

Figure 11. Yield of the molecular $C_{11}H_{12}N_2O_2^+$ tryptophan ion (a) and its ionic $C_9H_8N^+$ fragment (b) vs electron energy. Dots – experimental data, solid curves – results of the least-square fitting.

In the case of the aliphatic amino acid molecules, the lowest ionization energy (IE) is mostly related to the NH_2 nitrogen lone pair ionization. However, for aromatic amino acids, the lowest IE is obviously related to the ionization of the aromatic ring π molecular orbital having the ionization potential lower than that for the *n*-orbital of the nitrogen atom. The vertical ionization energies were calculated in [32, 33] for three conformations of tryptophan differing in the relative orientations of the amine and carboxyl heads. It is known that conformation can have a dramatic effect on the order of the ionized MOs and, thus, on the IEs [33].

Thus, for the three tryptophan conformers, the vertical IE calculated using the outer-valence-Green's-function (OVGF) method for the neutral molecule geometry lies within the 7.07–7.34 eV energy interval [33], and this agrees excellently with our calculated data shown in Table 1. For all the conformers the order of the first three MOs is the same and there are the π_1, π_2 and π_3 orbitals of the indole unity. The difference in the MO character is caused by the functional groups spatial orientation. The first ionization threshold value for the tryptophan molecule calculated using different methods (OVGF, Koopman's and MP2) is 7.22–7.88 eV [33].

According to our calculations, the ionization energy calculated as the difference in the energies between the tryptophan cation and the neutral molecule for conformer II is 7.10 eV, so this value coincides excellently with the data of [32].

Structurally, tryptophan, as the aromatic amino acid, could be presented as the indole-3-alanine or the 3-methyleneindole bound at the C_α position to a residue of a simple amino acid (i.e., glycine). The ion fragment $C_9H_8N^+$ (see Figure 11b), the methyleneindole, that corresponds to the dominant mass-spectrum peak, independent of the ionization path [11, 25, 26], is produced due to the C_α–C_β (C10-C11) bond rupture in the initial molecule. The intensity of the complementary ion with the mass m=74 a.m.u. (i.e., $C_2H_4NO_2^+$) is about 1% of that of the main peak. Thus, the tryptophan molecule ionization proceeds mainly due to the electron elimination by the methyleneindole group of the molecule at which the positive charge is predominantly localized:

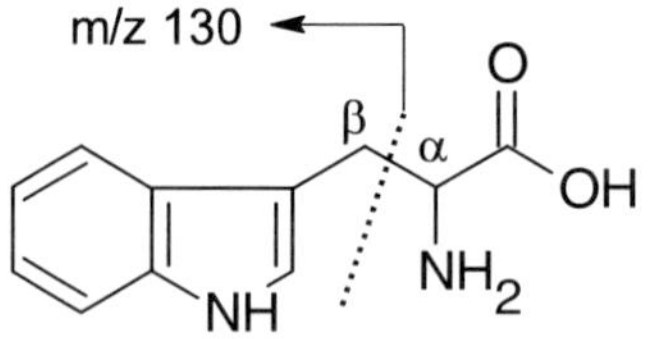

The bond break between the indole group and $-CH_2-C(NH_2)-COOH$ takes place as well.

However, this dissociation channel is not efficient: the intensity of the *m*=117 a.m.u. peak, corresponding to the indole molecule, ion does not exceed

4% of that of the main peak in the mass-spectrum. The dominant C10-C11 bond rupture as compared to the C4-C10 one is, in our opinion, because of the difference in the IEs of the fragments produced. The IE value for the 3-methylindole was determined in [34] to be 7.52 eV, while that for indole is 7.76 eV [35]. From the thermodynamic point of view, in case of the molecular ion dissociation, the charge is located on the fragment with a lower ionization potential and the fragmentation proceeds most probably via the simple bonds nearest to the charge localization places.

At the electron energies slightly exceeding the tryptophan molecule ionization energy, the probability of the C10–C11 bond dissociation is very high. At higher energies, when the electron energy is sufficient to ionize deeper molecular orbitals, first of all those of the nitrogen and oxygen lone pairs, this results in the dissociation of the neighboring bonds, thus, producing more fragmentation channels.

However, as seen from our mass-spectrum data, the signal intensity from fragments produced due to such channels is insignificant because the aromatic group stabilizes the positive charge and reduces fragmentation.

As mentioned earlier, the main peak in the tryptophan mass spectrum is due to the $C_9H_8N^+$ ion produced at a simple cleavage of the C_α–C_β (C10-C11) bond of the initial molecule. Here the charge is mainly localized on the fragment comprising the aromatic indole ring. The calculated binding energy per atom for the $C_9H_8N^+$ fragment is 5.44 eV and is higher than that of the complementary ion with the m=74 a.m.u. mass (3.92 eV). These results indicate a high stability of the m=130 a.m.u. fragment and a larger possibility of its formation. Hence, the peak at the m=74 a.m.u. mass is not intense in the tryptophan mass spectrum.

The inclusion of any possible cases of charge distribution at this bond dissociation results in the production of the C_9H_8N and $C_2H_4NO_2$ fragments according to the following pathways:

$$C_{11}H_{12}N_2O_2 + e \rightarrow \begin{cases} C_9H_8N^+ + C_2H_4NO_2^- + e & (1) \\ C_9H_8N^+ + C_2H_4NO_2^0 + 2e & (2) \\ C_9H_8N^- + C_2H_4NO_2^+ + e & (3) \\ C_9H_8N^0 + C_2H_4NO_2^+ + 2e & (4) \end{cases}$$

The lower stability of the $C_2H_4NO_2^+$ ion as compared to that of the $C_9H_8N^+$ fragment (see Table 3) and its higher appearance energy confirm the experimental results, i.e., the probability of the formation of a positively charged glycyl fragment is lower than that of the C_9H_8N one.

These results coincide with the observations of MacLennan et al [3] that indicate the C_α–C_β (C10-C11) bond cleavage with the elimination of the cationic side-chain fragment, $C_9H_8N^+$, as a low-energy process, while the production of $C_2H_4NO_2^+$ as a high-energy one.

Table 3. Calculated appearance energies (in eV) for the C_9H_8N and $C_2H_4NO_2$ fragments formed

C_9H_8N (m=130 a.m.u.) fragment charge	$C_2H_4NO_2$ (m=74 a.m.u.) fragment charge	Calculated appearance energy
1	-1	7.61
1	0	8.08
-1	1	8.99
0	1	9.13

According to the energy dependence of the $C_9H_8N^+$ ion (m=130 a.m.u.) yield from the parent tryptophan molecule (see Figure 11b), we have determined its appearance potential to be 9.1$\pm$0.1 eV and this value slightly (by 0.92 eV) exceeds the theoretical one obtained for pathway (2).

In the tryptophan mass-spectrum, the peak with the m=131 a.m.u mass is the second by its intensity, and the ratio of intensities of the ion peaks with the m=130 and m=131 a.m.u. masses at the 66 eV electron ionization energy shows, in our opinion, that the ion peak at m=131 a.m.u. is the first isotope peak. It should be noted that the intensity of this peak at photoionization by the noble-gas resonance radiation at the 21.2 eV energy, quoted in [26] (17.8% of the main peak intensity), exceeds considerably the calculated value of the first isotope peak intensity (10.3%).

Thus, we have analyzed two possibilities of the m=131 a.m.u. fragment formation due to the intramolecular hydrogen transfer, i.e., the cleavage of the C_α–C_β bond is accompanied by the hydrogen atom transfer from the O-H or N12-H23 bond, whereas the H atom is joined to the C_9H_8N fragment.

The above possibilities are chosen according to the results of the studies on the bond orders and bond lengths of the ionized and neutral molecules, i.e., the O-H bond becomes the weakest one in the ionized molecule, while the N12-H23 bond is one of the nearest to the weak bond because of the C_9H_8N

fragment. The calculated appearance energy indicates the pathway of formation of the *m*=131 a.m.u. fragment due to the intramolecular hydrogen transfer as very possible (see Tables 4 and 5).

Table 4. Calculated appearance energies (in eV) for the $C_2H_3NO_2$ and (C_9H_8N + H) fragments formed when the O-H bond is broken

$C_2H_3NO_2$ (*m*=73 a.m.u.) fragment charge	(C_9H_8N + H) (*m*=131 a.m.u.) fragment charge	Calculated appearance energy
-1	1	10.31
0	1	10.64

Table 5. Calculated appearance energies (in eV) for the $C_2H_3NO_2$ and (C_9H_8N + H) fragments formed when the N-H bond is broken

$C_2H_3NO_2$ (*m*=73 a.m.u.) fragment charge	(C_9H_8N +H) (*m*=131 a.m.u.) fragment charge	Calculated appearance energy
-1	1	14.31
0	1	12.02

Probably, at low ionizing radiation energy (21.2 eV), the intramolecular hydrogen transfer reaction is significant and the probability of the hydrogen atom migration from the hydroxyl group to the indole ring is larger than that for the amino group. However, at the 66 eV energy the peak at *m*=131 a.m.u. has the isotope character and this fact is indirectly confirmed by the presence of the peak at *m*=132. a.m.u. with the intensity of about 10% of that at *m*=131 a.m.u. corresponding to the calculated value for the second isotope peak $^{13}C_2C_7H_8N^+$.

In the tryptophan mass spectrum, the peaks with the *m*=103 and 77 a.m.u. masses demonstrate significant intensities and could be due to the initial molecule dissociation and result from the secondary fragmentation of the 3-methyleneindole ($C_9H_8N^+$, *m*=130 a.m.u.) molecule. The main channels of the dissociative ionization of the above molecule lead to the production of the ions with the *m*=103 a.m.u. (10% intensity) and the *m*=77 a.m.u. (14% intensity) masses [11].

The peak at the mass *m*=103 a.m.u. may correspond to the two isobaric ions with the gross-formula $C_8H_7^+$ and $C_7H_5N^+$. The C_8H_7 fragment with the vinyl-benzene structure is formed when the cleavage of the C_α–C_β bond is accompanied by a hydrogen atom transfer from the COOH group to the indole ring and the CH_2N (m=28 a.m.u.) fragment formation of the N9, C8 and H20, H21 atoms. The N9-C2 and C8-C7 bond elongation and the bond order value decrease in the ionized molecule (as compared to those of the neutral one) indicate that the destruction of the indole ring of the tryptophan molecule is possible (Table 2). The minimum energy-consuming pathway (see Table 6) of the C_8H_7 fragment appearance is as follows:

$$C_{11}H_{12}N_2O_2 + e \rightarrow (C_8H_6 + H)^+ + (CH_2N + C_2H_3NO_2)^- + e \rightarrow$$
$$C_8H_7^+ + (CH_2N + C_2H_3NO_2)^- + e \qquad (5)$$

Table 6. Calculated appearance energies (in eV) for the $C_8H_7^+$ and (CH_2N + $C_2H_3NO_2$) fragments formed

C_8H_7 (m=103 a.m.u.) fragment charge	($CH_2N + C_2H_3NO_2$) fragment charge	Calculated appearance energy
1	-1	14.53
1	0	18.49

We have calculated the appearance energy for the other isobaric fragment with the m=103 a.m.u. mass having the isocyanobenzene structure and produced via the following pathways:

$$C_{11}H_{12}N_2O_2 + e \rightarrow \begin{cases} C_7H_5N^+ + C_4H_7NO_2^- + e & (6) \\ C_7H_5N^+ + C_4H_7NO_2^0 + 2e & (7) \\ C_7H_5N^- + C_4H_7NO_2^+ + e & (8) \\ C_7H_5N^0 + C_4H_7NO_2^+ + 2e & (9) \end{cases}$$

Several structural isomers of this ion with different initial molecule bonds ruptured were investigated.

The appearance energy calculated for the C_7H_5N fragment with the isocyanobenzene structure is presented in Table 7 and indicates pathway (7) as the most probable one.

The comparison of the appearance energies calculated by us for the isobaric ions with the m=103 a.m.u. mass leads to the statement that the

$C_7H_5N^+$ fragment with the vinyl-benzene structure is produced in the near-threshold incident electron energy region. Above 12–15 eV, the dissociation of the C7-C8 and C7-C3 bonds and production of the C_8H_7 ion become possible.

Table 7. Calculated appearance energies (in eV) for the $C_7H_5N^+$ and (H + $C_4H_6NO_2$) fragments formed

C_7H_5N (*m*=103 a.m.u.) fragment charge	(H + $C_4H_6NO_2$) fragment charge	Calculated appearance Energy
1	-1	14.31
1	0	12.02

The ion peak with the *m*=77 a.m.u. mass could correspond to the two isobaric fragments, i.e., the $C_6H_5^+$ and $C_5H_3N^+$ ones. The formation of the latter fragment, taking into account the minimal number of the structural changes, is possible at the rupture of the three skeleton bonds, when two of them have the bond order more than 1. On the other hand, the possibility of rupturing the bonds with the order of more than 1 is very low. Thus, we assume that the $C_5H_3N^+$ fragment could be formed due to the indole ring destruction and the cleavage of the C_α–C_β bond. The calculated appearance energies for this ion are presented in Table 8.

The appearance energy value for this fragment confirms both the complicated way of its production and stepwise character of this process, i.e., the cleavage of the C_α–C_β bond is accompanied by the vinyl-acetylene molecule elimination.

Table 8. Calculated appearance energies (in eV) for the $C_5H_3N^+$ and (C_4H_4 + H + $C_2H_4NO_2$) fragments formed

C_5H_3N (*m*=77 a.m.u.) fragment charge	(C_4H_4 + H + $C_2H_4NO_2$) fragment charge	Calculated appearance energy
1	-1	26.52
1	0	20.56

The $C_6H_5^+$ ion is obviously the phenyl-radical that can be formed due to the dissociation of the C2-N9 and C3-C7 bonds of the tryptophan molecule followed by the hydrogen atom migration.

The calculated appearance energies for this ion, formed with minimal structural changes, are presented in Table 9.

Our calculations of the appearance energies for the isobaric fragments with m=77 a.m.u. show that the $C_6H_5^+$ ion is most likely to be produced below 20 eV, while above this energy, the C_5H_3N fragment could be formed.

Table 9. Calculated appearance energies (in eV) for the $C_6H_5^+$ and ($C_5H_7N_2O_2$) fragments formed

C_6H_5 (m=77 a.m.u.) fragment charge	$C_5H_7N_2O_2$ (m= 127 a.m.u.) fragment charge	Calculated appearance energy
1	-1	14.10
1	0	14.81

In the case of one hydrogen atom migration from the aliphatic chain to the indole moiety in the initial tryptophan molecule, the C7-C10 bond rupture leads to the production of the following complementary ions having the indole (C_8H_7N) and alanine ($C_3H_5NO_2$) structures with the ratio (5:1):

$$C_{11}H_{12}N_2O_2 + e \rightarrow \begin{cases} C_8H_7N^- + C_3H_5NO_2^{\ +} + e & (10) \\ C_8H_7N^0 + C_3H_5NO_2^{\ +} + 2e & (11) \\ C_8H_7N^+ + C_3H_5NO_2^{\ -} + e & (12) \\ C_8H_7N^+ + C_3H_5NO_2^{\ 0} + 2e & (13) \end{cases}$$

The calculated appearance energies for these fragments are shown in Table 10. For the indole ion they are somewhat lower than those for the ion with the $C_3H_5NO_2$ structure of dehydroalanine formed when the C7-C10 and C11-H25 bonds are broken.

Table 10. Calculated appearance energies (eV) for the C_8H_7N and $C_3H_5NO_2$ fragments formed

C_8H_7N (m=117 a.m.u.) fragment charge	$C_3H_5NO_2$ (m=87 a.m.u.) fragment charge	Calculated appearance energy
-1	1	9.84
0	1	8.63
1	-1	8.41
1	0	7.98

It should be pointed out that, according to our calculations, the appearance energy value for the $C_8H_7N^+$ ion produced by pathway (13) is 0.08 eV larger than the electron binding energy for the π_1-orbital of the indole group from the photoemission valence spectrum of tryptophan [26], i.e., the dissociation of the C7-C10 bond of the tryptophan molecule is possible at the high-order molecular orbital ionization. In this case, the dominant stabilization of the charge by the indole ring, as compared to the aliphatic amino acid residue, agrees fairly well with the ionization energies for the above tryptophan molecule components.

In the case of the hydrogen atom transfer in the transient ion from the indole ring to the aliphatic group, the dissociation of the C7-C10 group leads to the appearance of the complementary $C_3H_7NO_2$ and C_8H_5N fragments (see their appearance energies in Table 11).

Table 11. Calculated appearance energies (eV) for the C_8H_5N and $C_3H_7NO_2$ fragments formed

$C_3H_7NO_2$ (m=89 a.m.u.) fragment charge	C_8H_5N (m=115 a.m.u.) fragment charge	Calculated appearance energy
-1	1	14.67
0	1	13.32
1	-1	11.76
1	0	13.85

The ratio of the ion peak intensities corresponding to the fragments with 115–117 a.m.u. masses indicates a higher stability of the positively charged indole ion at the two hydrogen atoms loss as compared to the case when only one hydrogen atom is lost.

In the 25–30 a.m.u. parent molecule spectrum area, the most intense peak is that of the ion with the m=28 a.m.u. mass. Indeed, this fragment may be CH_2N and/or CO as observed in the other mass spectra of amine acids – alanine, methionine and glycine [28, 31, 37]. The formation of the m=28 a.m.u. fragment and the comparison of the stability of the positively charged CH_2N^+ and CO^+ ions indicate that the CH_2N^+ cation is more probable than the CO^+ one. Moreover, the comparison of the bond length and the bond order of the neutral and ionized tryptophan molecule proves the fact that the CH_2N fragment formation is more probable than that of the CO one. It should be mentioned that we calculated the appearance energy of the CH_2N^+ cation formation according to two possible pathways:

$$C_{11}H_{12}N_2O_2 + e \rightarrow CH_2N^+ + (C_{10}H_{10}NO_2)^{-/0} + ne \quad (14)$$

$$C_{11}H_{12}N_2O_2 + e \rightarrow CH_2N^+ + (COOH{+}C_9H_8N)^{-/0} + ne, \quad (15)$$

where *n* equals to 1 or 2 only. The calculated appearance energies for the *m*=28 a.m.u. fragment by two possible pathways (14) and (15) are presented in Tables 12 and 13.

Table 12. Calculated appearance energies (in eV) for the CH_2N^+ and $C_{10}H_{10}NO_2$ fragments formed

CH_2N (*m*=28 a.m.u.) fragment charge	$C_{10}H_{10}NO_2$ (*m* =176 a.m.u.) fragment charge	Calculated appearance energy
1	-1	12.83
1	0	15.48

Table 13. Calculated appearance energies (in eV) for the CH_2N^+ and ($COOH + C_9H_8N$) fragments formed

CH_2N (*m*=28 a.m.u.) fragment charge	($COOH + C_9H_8N$) fragment charge	Calculated appearance energy
1	-1	9.66
1	0	10.98

The analysis of the appearance energy for the *m*=28 a.m.u. fragment indicates that the channel with C11-N12-H_2 elimination is energetically preferable, although, at the energies above 15.5 eV, the destruction of the indole ring due to electron impact followed by the CH_2N^+ ion formation is possible.

On the other hand, it should be mentioned that, according to pathway (5), the CH_2N fragment (*m*=28 a.m.u.) is also produced. There is a very low probability for the CH_2N fragment to be charged positively in the (CH_2N + $C_2H_3NO_2$)$^-$ compound because $C_2H_3NO_2^{-2}$ could be formed in this case. Besides, CH_2N^+ could be formed during the secondary fragmentation of the *m*=130 a.m.u. fragment ($C_9H_8N^+$). In this case, the minimal calculated energy required for this secondary dissociation is 5.84 eV, while depending on the fragment charge distribution the total energy required for this two-stage process is 13.43–13.91 eV,.

In the tryptophan molecule mass-spectrum, the peaks with the fractional masses corresponding to the doubly charged ions are revealed. The most intense peak corresponds to the $C_{10}H_{11}N_2^{2+}$ ion and results from the carboxyl group detachment from the parent molecule. The intensity of this doubly charged ion peak is approximately the same as that for the relevant singly charged ion. At the protonated tryptophan ionization under the single-collision conditions [38], the yield curve for the ion with the *m*=159 a.m.u. mass has its maximum, and this testifies to the predisposition of this ion to the secondary dissociation. Probably, the detachment of the carboxyl group leads to the formation of a conjugated bond system in the aliphatic part of the molecule, favoring the charge stabilization by this fragment.

The low intensity peaks with the *m*=57.5, 64.5 and 65.5 a.m.u. masses correspond to the $C_8H_5N^{2+}$, $C_9H_7N^{2+}$ and $C_9H_9N^{2+}$ ions, respectively, and demonstrate the indole ring ability to keep stability when detaching the different number of the hydrogen atoms with the simultaneous elimination of two electrons.

The $C_8H_5N^{2+}$ fragment could be formed from the neutral tryptophan due to the C7-C10 bond destruction and the simultaneous H atom migration from the $C_8H_6N^+$ fragment. We calculated the appearance energy of this fragment for both cases. The comparison of the appearance energies obtained proves the fact that formation of the $C_8H_5N^{2+}$ ion according to the following pathway requires the minimal energy:

$$C_8H_6N^+ - e \rightarrow C_8H_5N^{2+} + H^0 \qquad (16)$$

Referring to the results on the appearance energy calculation, we predict the two possible ways of $C_9H_7N^{2+}$ formation according to the following schemes:

$$C_{11}H_{12}N_2O_2 + e \rightarrow C_9H_7N^{2+} + C_2H_5NO_2^{\,0} + 3e, \qquad (17)$$

where the C10-C11 bond destruction is accompanied by the H26 atom migration producing a stable glycine molecule and

$$C_{11}H_{12}N_2O_2 + e \rightarrow C_9H_8N^{+} + C_2H_4NO_2^{-} + e \qquad (18)$$

$$\searrow C_9H_7N^{2+} + H^0 + e$$

with the hydrogen atom detachment from the doubly charged fragment as a result of the secondary dissociation.

The calculated appearance energy for the $C_9H_7N^{2+}$ fragment is equal to 22.99 eV in the case of pathway (17), while that in the case of pathway (18) is not less than 25.91 eV. Hence, pathway (17) is more energetically favorable to produce the doubly charged $C_9H_7N^{2+}$ ion. The $C_9H_9N^{2+}$ ion in this case could be formed as follows:

$$C_{11}H_{12}N_2O_2 + e \rightarrow C_9H_9N^{2+} + (C_2H_3NO_2\ )^{-/0} + ne\ (n=2,\ 3) \qquad (19)$$

The iminoacetic acid is formed as the complementary fragment, and the appearance energy for the doubly charged $C_9H_9N^{2+}$ ion is approximately 21.59 eV and 21.32 eV for the negatively charged one and the neutral iminoacetic acid, respectively.

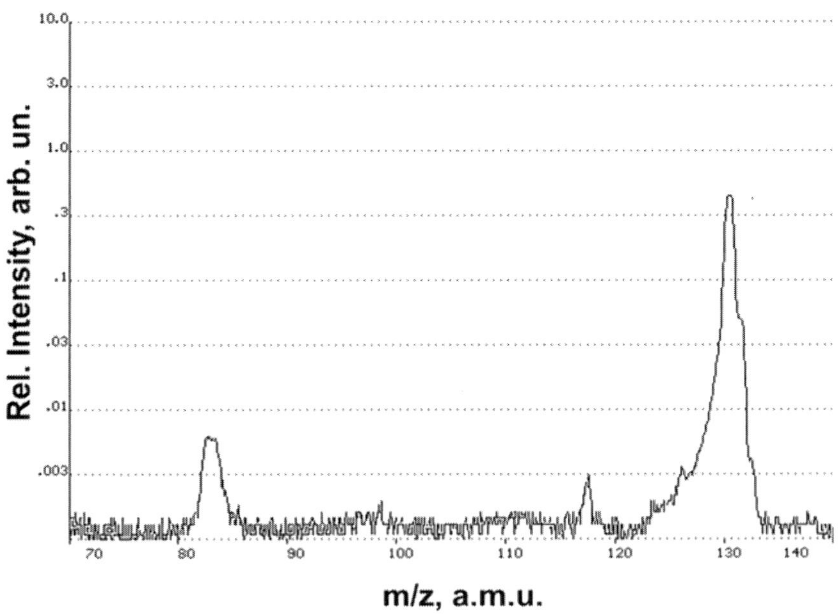

Figure 12. The area of the tryptophan molecule mass-spectrum at the 18 eV electron energy.

The presence of the peak at about m=81.5 a.m.u. at the ionizing electron energy of 18 eV (see Figure 12) shows that this peak does not correspond to the doubly charged ion, while the fuzzy shape of this peak indicates the presence of a metastable ion with the lifetime less than 10^{-6} s, so it decays on its path from the ion source to the detector. The estimation of the masses of the initial (m_1) and resulting (m_2) ions using formula $m^*=m_2^2/m_1$, where m^* is the diffuse peak mass, shows that the peak at m~81.5 a.m.u. corresponds to the decay of the ion with the m=130 a.m.u. mass and the production of the ion with the m=103 a.m.u. mass.

Since our mass-spectrum reveals the peaks with the m=103 and 130 a.m.u. masses and the diffuse peak with the m~81,5 a.m.u. mass, one may conclude that at least a part of the $C_8H_7^+$ ions is produced directly from the $C_9H_8N^+$ ion. Thus, at the internal energy excess, the methyleneindole ion decays producing the $C_8H_7^+$ ion.

High-Energy Electron Impact Fragmentation

The understanding of the radiation chemistry of amino acids is substantially important with respect to realizing how radiation interacts with proteins, the amino acid molecules being their elementary building blocks [2, 3]. In spite of very intensive efforts of a great number of researchers, the radiation chemistry of amino acids in the solid state remains actual and requires further clarification.

Systematic studies of the stability of amino acids under the influence of high energy radiation show that almost in all cases they can withstand an enormous radiation dose if applied in the solid state. The continuous radiolysis causes a steady decomposition, but occurring in the solid state, the radiolysis products are not so mobile and cannot escape easily, especially, large molecular fragments remain trapped in the crystalline structure. The presence of the side chains in amino acids generally modifies the radical formation processes in a way specific for each side chain but, in general, aromatic molecules are more radiation resistant than the aliphatic ones [39].

Depending upon the kinetic energy of the electron, the geminate recombination may be avoided and cause several ionizations and excitations before it eventually becomes trapped at different molecules distantly from the site of the initial trace.

The consequences of excitations are of minor importance since the molecules mainly return to their ground states due to radiative or non-radiative relaxations.

Thus, a high-energy radiation impact results in the radical anions as the reduced species and radical cations as the oxidized ones [40].

Decarboxylation is one of the two primary destruction mechanisms in irradiated amino acids. This pathway starts with one-electron oxidation followed by the H^+ transfer and, finally, the H-atom detachment from a neighboring molecule aiming to produce an amine and the CO_2 molecule [40]:

$$^{+}NH_3CH(R)COO^{-} \rightarrow NH_2CH_2(R) + CO_2 \quad (20)$$

The other major amino acid destruction route is reductive deamination, which produces ammonia and a free radical:

$$^{+}NH_3CH(R)COO^{-} + H \rightarrow NH_3 + {}^{\bullet}CH(R)COOH \quad (21)$$

The above radical can extract a hydrogen atom to make carboxylic acid [41]. The irradiation of L-tryptophan in the solid state to a dose of 3.2 MGy leads to the reduction of ~8.1°C in the melting point peak due to the formation of non-volatile radiolysis products, which remain trapped in the crystals of tryptophan. The residual amount of the pristine molecules after such irradiation, estimated by the differential scanning calorimetry technique, is 92% [42]. Thus, after such a high deposited dose, most of the tryptophan molecules remain intact.

As mentioned above, we have measured the mass spectrum of the tryptophan molecule after its interaction with a beam of electrons of the 9.5 MeV energy generated by an M30 microtron. The sample irradiated by the electron accelerator was studied as described above, and the obtained mass spectrum was compared to that of the initial non-irradiated one (see Figure 13). For the correct comparison, the intensities of the parent ion peaks in the two spectra were reduced to the same value.

First, it should be noted that the mass spectrum of the tryptophan molecule irradiated by the 9.5 MeV energy electron beam is generally similar to that of the non-irradiated molecule. However, in several cases the intensities of some peaks (say, with the m=28 a.m.u. and m=77 a.m.u. masses) significantly differ from each other. The intensities of these two peaks for the irradiated molecule are 4 times and more than 1.5 times, respectively, larger than those for the non-irradiated molecule.

It is interesting that in our mass-spectrometric studies the abrupt decrease in the intensity of the low-mass (m=16–18, 27–30, 44–46 a.m.u.) ions in the mass-spectra was noticed at the initial stage of the measurements. In our opinion, this testifies to the fact of the presence of the volatile products in the irradiated tryptophan powder adsorbed by the crystalline lattice. The most probable volatile radiolysis products are NH_3, H_2O, CO_2 and CO_2H_2 [43, 44], while other products remain trapped in the damaged crystal structure together with the residual amount of non degraded amino acid.

The results of the theoretical analysis exhibit the difference less than 0.02 eV (see Table 1) in the total energies of the most stable conformers, indicating that the possibilities of production of these conformers are equal. On the other hand, this observation allows us to predict that one of these molecule conformers could appear and became stable in the crystalline lattice due to irradiation, when the alanine ($C_3H_6NO_2$) part of the tryptophan molecule is rotated with respect to its indole part and could be used to prove the effects caused by persistent changes due to irradiation.

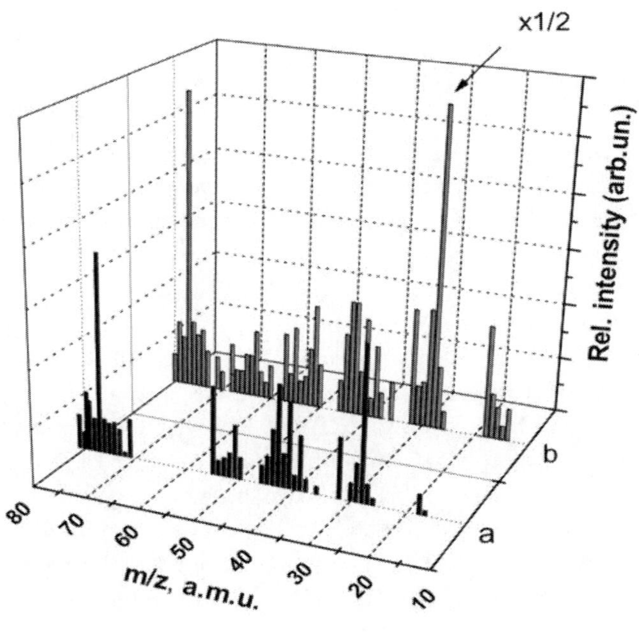

Figure 13. Mass spectra of both non-irradiated (a) and irradiated (b) tryptophan molecules.

As mentioned above, the C10-C11 bond (see Figure 8) cleavage leads to the formation of the most abundant fragment with m=130 a.m.u. and the gross formula $C_9H_8N^+$. This implies more probable cleavage of the above bond for conformer I and allows us to assume a reason why the intensities of some peaks in the mass spectra of both non-irradiated and irradiated molecules are different, although these spectra are quite similar.

The most significant is the difference in the intensities of the three main peaks in the mass spectra of irradiated and non-irradiated tryptophan, i.e., those with m=103 a.m.u., m=77 a.m.u., and m=28 a.m.u., as compared to that of the most abundant fragment with m=130 a.m.u. Referring to the results obtained, we suggest that the positively charged fragment with the m=103 a.m.u. mass could be $C_8H_7^+$ and/or $C_7H_5N^+$ and its intensity becomes twice lower after irradiation, while those at m=77 a.m.u. (i.e., $C_5H_3N^+$ and/or $C_6H_5^+$) and m=28 a.m.u. (CH_2N^+) increased in intensity. The formation of these fragments is related to the cleavage of different bonds of the molecule followed by the H atoms migration.

The decrease in the intensity of the peak at m=103 a.m.u. with respect to the main peak in the mass spectrum could indicate the reduction of the efficiency of one of the three channels of the initial molecule decay (pathways (5), (7)) and the cascade 130→103 mechanism for the peak at m~81.5 a.m.u.) after the irradiation. As to the ions with m=77 a.m.u. and 28 a.m.u., we can assume different possibilities of their formation: from the intact molecule as for the non-irradiated case, and from long-lived radicals and radical ions such as in alanine [45] after the irradiation. As for the theoretical calculations of the appearance energies for the fragment ions at the radical dissociation, they are not completed at the moment. In addition, the increase in the intensity of the low-mass ions could also result from the decay of a part of the tryptophan molecules under irradiation that remain trapped in the damaged crystal structure together with the residual amount of not degraded amino acid.

Referring to the results presented in [8], there is no doubt that the peak with m=28 a.m.u. in both spectra is CH_2N^+. Hence, the increase of the peak intensity could not be explained as the formation of other fragments with the m=28 a.m.u. mass. However, there are two possible ways of this fragment formation: i) the CH_2N and $C_{10}H_{10}NO_2$ fragments are produced as the final products of the fragmentation (i.e., due to the fragmentation of the indole part of tryptophan); ii) the CH_2N and COOH, H, C_9H_8N fragments are produced (i.e., due to fragmentation of the alanine part of tryptophan, see pathways (14, 15)). It should be mentioned that theoretical results exhibit an approximately 3 eV difference in the appearance energies of the m=28 a.m.u. fragment for conformer II obtained by two ways described above. Moreover, the appearance energies for this fragment originated from different conformers due to the fragmentation of the same part of tryptophan differ slightly as well. As example, the appearance energy of the fragment with m=28 a.m.u., when the indole part is destructed, is equal to 9.6 eV and 9.7 eV for conformers I and II, respectively. These observations allow us to predict different possibilities of realization of the above fragmentation ways for different conformers. Hence, the effects caused by the persistent changes due to irradiation could lead to the appearance of the same chemical composition fragments obtained by different ways and this could be a reason why the intensity of the peak at m=28 a.m.u. in the spectrum of the irradiated molecule is four times larger than that of the non- irradiated molecule.

A slight contribution to the m=28 a.m.u. peak intensity may also be due to the residual amount of CO formed at the radiolysis of a part of tryptophan molecules. However, in view of a high volatility of this compound and a small fraction of molecules damaged by radiation [42], we may neglect this ion contribution into the m=28 a.m.u. peak intensity.

Hence, the results of our experimental and theoretical investigations exhibit the feasible persistent change of the tryptophan structure due to irradiation leading to the increase of some peaks intensities in the mass spectra because of the appearance of the fragments either with the same mass but different chemical composition and/or that of the fragments with the same mass and chemical composition but obtained by different ways.

ACKNOWLEDGMENTS

The authors would like to thank I.G. Megela from the Institute of Electron Physics (Uzhgorod, Ukraine), Prof. M. Cegla from the Jagellonian University (Krakow, Poland) and V. Patasiene from Vilnius University (Vilnius, Lithuania) for their cordial assistance and fruitful discussions. Special thanks go to the Digital Science and Computing Center of Vilnius University for the resources and technical support provided.

REFERENCES

[1] H. Abdoul-Carime, S. Gohlke and E. Illenberger, *Chem. Phys. Let.* 402, 497 (2005).

[2] A. F. Fuciarelli and J. D. Zimbrick, *Radiation Damage in DNA: Structure/Function Relationship at Early Times*, (Battelle Press, Columbus OH, 1995).

[3] M. S. MacLennan, K. N. Sutherland and G. Orlova, *Journ. Mol. Struct.: Theochem* 822, 21 (2007).

[4] P. K. Pallaghy, A. P. Melnikova, E.C. Jimenez, B.M. Olivera and R.S. Norton, *Biochemistry* 38 (35), 11553 (1999).

[5] L. V. Hankes, R. R. Brown, J. Leklem, M. Schmaeler and J. Jesseph, *J. Invest. Dermatol.* 58, 85 (1972).

[6] J. Thomson, H. Rankin, G.W. Ashcroft, C.M. Yates, J.K. McQueen and S. W. Cummings, *Psychological Medicine* 12 (4), 741 (1982).

[7] H. Kang, C. Dedonder-Lardeux, C. Jouvet, S. Martrenchard, G. Grégoire, C. Desfrançois, J.-P. Schermann, M. Barat and J.A. Fayeton, Phys. Chem. Chem. Phys. 6, 2628 (2004).

[8] V. S. Vukstich, A .I. Imre, and A. V. Snegursky, *Instr. Exper. Tech.* 54, 66 (2011).

[9] A.V. Snegursky, J. Tamuliene, V.S. Vukstich and L.G. Romanova, Methionine Molecule Electron-Impact-Induced Fragmentation: Mechanisms and Chemical Structure. In: *Methionine, Biosynthesis, Chemical Structure and Toxicity.* – New York: Nova Science Publishers (ed. A.V. Snegursky). (2013), 187 pp.

[10] G. Hanel, B. Gstir, T. Fiegele, F. Hagelberg, K. Becker, P. Scheier and A. Snegursky, *J.Chem. Phys.* 116, 2456 (2002).

[11] National Institute of Standards, Standard Reference Database: *Chemistry Webbook* http://webbook.nist gov.

[12] MI 1201 *Mass Spectrometer. Instruction Manual, SELMI Corporation,* (2002).

[13] *NuDAQ. ACL-8112 Series Enhanced Multi-Functions Data Acquisition Cards. Instruction Manual.* (2011).

[14] A. D. Becke, *J. Chem. Phys.* 98, 5648 (1993).

[15] R. A. Kendall, T.H. Dunning Jr. and R.J. Harrison, *J. Chem. Phys.* 96, 6796 (1992).

[16] J. T. Bursey, M.M. Bursey and D.G.I. Kingston, *Chem. Rev.* 73, 231 (1973).

[17] *Gaussian 03, Revision C.02*, *Gaussian, Inc., Wallingford CT*, (2004).

[18] O. V. Kharitonova, *Methods and Technology of Production of Organic Compounds Used in Materials for Information Registration.* Part I, (MITHT Publishers, Moscow, 2004). (in Russian).

[19] V. V. Tunina, *Sublimation enthalpy and structural characteristics of some amino acids and dipeptides: experiment and quantum-chemical calculations.* Ph.D. Thesis. (Ivanovo, 2014). 134 pp. (in Russian).

[20] A. Kaczor, I. D. Reva, L. M. Proniewicz, and R. Fausto, *J. Phys. Chem. A*, 111, 5692 (2007).

[21] W. Zhang, V. Carravetta, O. Plekan, V. Feyer, R. Richter, M. Coreno, and K.C. Prince, J. Chem. Phys. 131, 035103 (2009).

[22] A. Kraj, D.M. Desiderio, N.M. Nibbering, *Mass Spectrometry: Instrumentation, Interpretation, and Applications*, edited by R. Ekman, J. Silberring, A. Westman-Brinkmalm, (John Wiley & Sons, 2009).

[23] R. S. Mulliken, J. Chem. Phys. 23, 1833 (1955).

[24] H. Jurgen, *Gross Mass Spectroscopy. A Textbook.* 2nd edn. (Springer-Verlag, Berlin-Heidelberg, 2011).

[25] F. O. Talbot, T. Tibault, A. Rodolphe, M. Broyer and P. Dugourd, *J. Chem. Phys*. 122, 074310 (2005).

[26] O. Plekan, V. Feyer, R. Richter, M. Coreno and K.C. Prince, *Mol. Phys*. 106, 1143 (2008).

[27] V. S. Vukstich, A.I. Imre, L.G. Romanova and A.V. Snegursky. *J. Phys. B* 43, 185208 (2010).

[28] J. Tamuliene, L.G. Romanova, V.S. Vukstich and A.V. Snegursky, *Chem. Phys*. 404, 74 (2012).

[29] V. S. Vukstich, L.G. Romanova and A.V. Snegursky, *Tech. Phys. Let.* 38, 347 (2012).

[30] J. Tamuliene, L.G. Romanova, V.S. Vukstich and A.V. Snegursky, *Chem. Phys*. 404, 36 (2012).

[31] P. H. Cannington and N.S. Ham, *J. Electron Spectrosc. Relat. Phenom.* 32, 139 (1983).

[32] V. Lepere, B. Lucas, M. Barat, J.A. Fayeton, V.J. Picard, C.Jouvet, P. Carcabal, I. Nielsen, C. Dedonder-Lardeux, G. Gregoire and A. Fujii, *J. Chem. Phys*. 127, 134313 (2007).

[33] D. Dehareng and G. Dive, *Int. J. Mol. Sci.* 5, 301 (2004).

[34] D. Q. Liu and M. Sun, *ISRN Spectroscopy*. 2012, 1 (2012).

[35] J. Hager, M. Ivanco, M.A. Smith and S.C. Wallace, *Chem. Phys.* 105, 397 (1986).

[36] V. S. Vukstich, L. G. Romanova, I. G. Megela and A. V. Snegursky, *Tech. Phys. Let.* 40, 263 (2014).

[37] J. Tamuliene, L. G. Romanova, V. S. Vukstich and A. V. Snegursky, *Lith. J. Phys.* 53, 195 (2013).

[38] H. El Aribi, G. Orlova, A. C. Hopkinson and K. W. M. Siu, *J. Phys. Chem. A.* 108, 3844 (2004).

[39] F. Cataldo, O. Ursini, G. Angelini, S. Iglesias-Groth and A. Manchado, *Rend. Fis. Acc. Lincei*. 22, 81 (2011).

[40] E. Sagstuenl, A. Sanderud, and E. O. Hole, *Radiat. Res.* 162, 112 (2004).

[41] K. Perry, A. Gerakines, R. L. Hudson, M. H. Moore and J.-L. Bell, *Icarus*, 220, 647 (2012).
[42] F. Cataldo, G. Angelini, S. Iglesias-Groth and A. Manchado, *Rad. Phys. Chem.* 80, 57 (2011).
[43] W. Kulik and W. Heerma, *Biomed. Envir. Mass Spectr.* 15, 419427 (1988).
[44] G. Kminek and J. L. Bada, *Earth and Planet Sci. Let.* 245, 1 (2006).
[45] E. Malinen, E. A. Hult, E. O. Holea and E Sagstuena, *Radiat. Res.* 159, 149 (2003).

In: New Developments in Tryptophan Research
Editor: Victoria Hayes
ISBN: 978-1-63483-025-6

Chapter 6

A NOVEL TRYPTOPHAN-SELECTIVE AMINO ACID TRANSPORTER THAT IS FUNCTIONALLY COUPLED TO IDO1-DEPENDENT SIGNALING PATHWAYS

Yangzom D. Bhutia and Vadivel Ganapathy
Department of Cell Biology and Biochemistry, Texas Tech University Health Sciences Center, Lubbock, TX, US

ABSTRACT

Tryptophan is not only an essential amino acid but also an effective signaling molecule as an indicator of amino acid sufficiency. It is catabolized by indoleamine 2,3-dioxygenase 1 (IDO1) with kynurenine as one of the final end products. Induction of IDO1 in cells by a variety of extracellular signals (e.g., interferon-γ, IFN-γ) leads to tryptophan depletion and kynurenine generation inside the cells. IDO1-mediated tryptophan catabolism impacts on three different effector pathways: activation of the protein kinase GCN2 that blocks the protein translation factor eIF-2 thereby blocking protein synthesis, suppression of mTOR thereby interfering with amino acid sufficiency signaling, and activation of the nuclear receptor AhR through kynurenine. As such, IDO1-dependent tryptophan metabolism influences a plethora of signaling pathways. IDO1 induced in one cell type can have an impact on a neighboring cell type through rapid degradation of tryptophan in IDO1-positive cells thereby causing depletion of extracellular tryptophan, thus making this important amino acid unavailable for the neighboring cell type. This is the mechanism responsible for the ability of IDO1-expressing antigen-presenting immune cells to suppress proliferation of effector T cells. An often-overlooked component of this mechanism is the transport process in the plasma membrane of IDO1-positive cells that delivers extracellular tryptophan to the intracellular IDO1. The existence of a transporter selective for tryptophan has been identified in IDO1-expressing cells at the functional level. As expected from the obligatory nature of the transporter for the ability of IDO1 to deplete extracellular tryptophan, the induction of the enzyme and the transporter occurs in a coordinated manner, in most cases, by the same signaling pathways. For example, IFN-γ and AhR that induce IDO1 also induce the tryptophan-selective transporter in the same cell. It is readily apparent that pharmacological blockade of IDO1-mediated tryptophan depletion has therapeutic potential in clinical and pathological conditions that require activation of the immune function as a treatment. Inhibition of either IDO1 or the tryptophan transporter would block this pathway, and both strategies might possibly have comparable therapeutic outcome. Nonetheless, current efforts in this area are almost exclusively focused on IDO1. One of the primary reasons for the apparent lack of interest in the tryptophan transporter as a therapeutic target is that neither the protein nor the gene responsible for the transport process has been identified at the molecular level. But pharmacological targeting of the transporter does not necessarily have to depend on the molecular identity of the transporter. The transport process is known at the functional level; IFNγ-inducible transport function that is selective for tryptophan can be used as a read out for the IDO1-associated tryptophan transporter in the design and development of pharmacological agents to block the transport process. Of course, information on the structure of the transporter protein would facilitate the discovery process in terms of modeling potential inhibitors or blockers of the transporter.

INTRODUCTION

Tryptophan is an essential amino acid that has to be provided in the diet because mammalian cells cannot synthesize it de novo. As with any amino acid (essential or non-essential), deficiency of tryptophan would lead to decreased protein synthesis in cells. There are signaling pathways operative in mammalian cells that detect amino acid sufficiency and transmit regulatory signals to protein synthesizing machinery; decreased amino acid levels leading to decreased protein synthesis and increased amino acid levels to increased protein synthesis. In addition to its role in protein synthesis, tryptophan participates in other important biological functions. The role of tryptophan as the precursor of the neurotransmitter serotonin and the hormone melatonin is widely known as evident from growing interest in this amino acid as a nutritional supplement for various ailments including depression and sleep disorders. But another important function of tryptophan as a modulator of immune system has come to the forefront in recent years; this is related to the enzyme indoleamine 2,3-dioxygenase 1 (IDO1) that catabolizes tryptophan. IDO1 is an inducible enzyme expressed in certain specific immune cell types including macrophages and antigen-presenting cells such as dendritic cells. Tryptophan breakdown by IDO1 in these cells has two important signaling components: (a) depletion of this essential amino acid both inside the cells and in the extracellular milieu in the immediate vicinity of the IDO1-expressing cells and (b) the breakdown product kynurenine serving as an endogenous agonist for aryl hydrocarbon receptor (AhR), a nuclear receptor functioning as a transcription factor. Sufficiency or deficiency of tryptophan inside the cells impacts on two important regulatory protein kinases controlling protein synthesis and cell growth and proliferation: GCN2 (General Control Nonderepressing-2) and mTOR (mechanistic Target Of Rapamycin). Similarly, availability of tryptophan in the extracellular milieu determines the intracellular level of this amino acid via the activity of amino acid transporters in the plasma membrane. Mammalian cells express several transporters to mediate cellular uptake of tryptophan, and these transporters are expressed differentially in different cell types, including immune cells. One particular transporter, almost specific only for tryptophan, is expressed in IDO1-expressing cells, indicating a functional link between the cellular uptake of tryptophan via the transporter and the intracellular breakdown of tryptophan via IDO1 to impact on intracellular as well as extracellular levels of this amino acid. AhR, the receptor for the tryptophan breakdown product kynurenine, on the other hand elicits dual but opposing effects on IDO1; a genomic effect contributing to induction of IDO1 and a non-genomic effect promoting proteasomal degradation of IDO1. In addition, AhR regulates other important signaling pathways involving various cytokines. Given the profound impact on the immune system, the tryptophan-selective transporter/IDO1/AhR pathway offers ideal druggable targets for suppression or activation of immune function as appropriate for a given clinical and pathological condition.

INDOLEAMINE 2,3-DIOXYGENASE 1 (IDO1)

Three distinct enzymes mediate the degradation of tryptophan: tryptophan 2,3-dioxygenase (TDO), IDO1, and IDO2 (1). All three enzymes contain heme at the catalytic site. TDO is expressed mostly in the liver and has little relevance to immune function. IDOs are expressed in a variety of cell types. IDO1 is particularly expressed in the placenta, colonic epithelial cells, macrophages and dendritic cells; it is induced by the proinflammatory cytokine interferon-γ (IFN-γ) and immune stimulants such as bacterial lipopolysaccharides. The role of IDO1 as a tryptophan-degrading enzyme has been known for a long time, but its role as a suppressor of immune function was discovered when its inhibition with 1-methyltryptophan was shown to induce rejection of semiallogeneic, but not syngeneic, fetuses in mice (2). Since then, IDO1 has been in the forefront in the field of immunology as a therapeutic target to modulate immune homeostasis (suppression or activation) in diverse physiological and pathological conditions including bacterial, fungal and viral infection, allergy, autoimmunity, organ transplantation and cancer (3, 4). IDO2 has a narrower tissue expression pattern than IDO1, but does function in immunomodulation and autoimmunity and seems to have some unique functions not shared by IDO1 (5).

IDO1 acts on tryptophan breaking the bond between carbons 2 and 3 in the indole ring by oxidative cleavage, thus generating formylkynurenine, which is then converted into kynurenine by kynurenine formamidase. Under inflammatory conditions, IDO1 is induced in dendritic cells and macrophages by IFN-γ, which is primarily generated by NK cells, and $CD4^+$ and CD^+ T cells. The rapid induction of IDO1 and consequent degradation of tryptophan results in tryptophan depletion in dendritic cells and in the extracellular milieu surrounding dendritic cells. As such, cytotoxic T cells in the immediate vicinity of dendritic cells are deprived of this essential amino acid, thus preventing their proliferation and hence suppressing immune responses. This phenomenon is believed to occur to mitigate immune-mediated tissue damage and curtail undue overactive immune responses under inflammatory conditions. T cells that come in contact with IDO1-expressing dendritic cells are now exposed to an extracellular milieu deficient in tryptophan at the contact sites (immunological synapse). As a result, T cells fail to proliferate, thus causing immune suppression. This concept of immune suppression by tryptophan depletion involves altered activities of GCN2 and mTOR in T cells (6). Decreased intracellular levels of tryptophan in T cells activates the protein kinase GCN2 through binding of unloaded Trp-tRNA to the kinase, which then phosphorylates eIF2α (eukaryotic initiation factor 2α); phospho-eIF2α reduces global protein synthesis, thereby suppressing proliferation of T cells and causing T cell anergy (a state of unresponsiveness to immune signals) (7). IDO1 is not only enzymatically active to degrade tryptophan but also functions as a signal-transducing molecule (8, 9). IDO1-expressing dendritic cells release type I interferons (IFN-α and IFN-β) that work in an autocrine and paracrine manner to maintain IDO1 induction and function and to generate TGF-β. IDO1 has two phosphorylatable immnoreceptor tyrosine-based inhibitory motifs (ITIMs), which undergo Fyn-mediated phosphorylation in dendritic cells in the presence of TGF-β and serve as docking sites for Src homology 2 domain phosphotyrosine phosphatases 1 and 2 (SHP-1 and SHP-2) and also for suppressor of cytokine signaling 3 (SOCS3). The binding of SHP-1 and SHP-2 to IDO1 activates a variety of downstream signaling pathways including noncanonical NF-κB, which converts dendritic cells into a regulatory phenotype and maintains the generation of type I IFNs and TGF-β that have profound impact on expansion and function of $FoxP3^+$ regulatory T cells (Tregs) (10). The regulatory dendritic cells also have the ability to convert naïve T cells into Tregs. The binding of SOCS3 to IDO1 recruits E3 ubiquitin ligase, thus bringing IDO1 close to the ubiquitination complex for subsequent ubiquitination and proteasomal degradation (11).

AhR as an Effector Pathway Downstream of IDO1: The IDO1 Metabolite Kynurenine as an Agonist for AhR

The IDO1 metabolite kynurenine also plays an obligatory role in IDO1-dependent immune suppression. Kynurenine is an endogenous agonist for AhR (12, 13). Activation of AhR in IDO1-expressing dendritic cells has at least three consequences: (a) it induces the expression of IDO1 and other AhR-responsive genes such as SOCS3 and IL-6 through AhR-responsive elements located on the respective gene promoters, (b) it leads to phosphorylation of IDO1 and subsequent production of TGF-β, and (c) it potentiates the ability of dendritic cells to convert naïve T cells into Tregs. By a non-genomic mechanism, the ligand-bound AhR assembles the ubiquitination complex containing the ubiquitin ligase CUL4B, which leads to ubiquitination of IDO1 and subsequent proteasomal degradation. Thus, the kynurenine-AhR axis is involved both in positive and in negative regulation of IDO1; while the genomic action of activated AhR could be considered as a positive feedback mechanism, the non-genomic action could be considered as a negative feedback mechanism. Both pathways are needed for exquisite control of IDO1 in immune cells.

mTOR As Another Effector Pathway Downstream of IDO1

The mTOR complex 1 (mTORC1) is a master regulator that coordinates diverse extracellular and intracellular signals to control metabolic pathways in cells (14). The extracellular signals include growth factors such as insulin, IGF-1 and EGF; the intracellular signals include cellular energy status (AMP/ATP ratio) and cellular nutritional sufficiency. mTORC1 is activated when cells are exposed to growth factors, energy surplus (i.e., decreased AMP/ATP ratio) and nutritional surplus. Amino acids represent an important class of nutrients that control mTOR pathway. While excess amino acids inside the cells activate mTORC1, amino acid deficiency suppresses mTORC1. The activation of mTORC1 by amino acids is versatile and complex (15). Specific amino acids such as leucine, glutamine, and arginine have been shown to participate in the process. Activation of mTORC1 by amino acids involves the assembly of the complex either on the lysosomal membrane or Golgi membrane. IDO1-mediated tryptophan deficiency in IDO1-positive cells inhibits mTORC1 through the amino acid-sensing protein kinase GLK1; presumably GLK1 activates mTORC1 and tryptophan deficiency blocks GLK1 (16, 17). Active mTORC1 has multiple effects on cell growth and proliferation driving anabolic pathways and suppressing catabolic pathways. IDO1-mediated blockade of mTORC1 in dendritic cells converts them into regulatory dendritic cells and also potentiates their ability to convert naïve T cells into Tregs. In addition, depletion of tryptophan in the immunological synapse between dendritic cells and T cells makes the latter cells tryptophan deficient, thus suppressing mTORC1 and mTORC1-dependent anabolic pathways, which leads to T cell anergy. This phenomenon is also associated with increased autophagy.

The Tryptophan-AhR Pathway in Host-Bacteria Symbiosis in Colon

The IDO1-dependent tryptophan catabolism and the subsequent involvement of AhR and mTORC1 in immune cells lead to a tolerogenic phenotype of the immune system. This includes conversion of dendritic cells into regulatory dendritic cells, suppression of cytotoxic T cells, conversion of naïve T cells into immunosuppressive Tregs, and activation of Tregs and myeloid-derived suppressor cells. The relevance of this phenomenon to the ability of tumor cells to evade the host immune system has received wide attention. IDO1 is upregulated in tumor cells as well as in immune cells in tumor-draining lymph nodes to exploit the tryptophan-AhR pathway of immunosuppression to their own advantage. This pathway also plays a critical role in certain physiological processes. A tolerogenic phenotype of the host immune system is obligatory to maintain the host-bacteria symbiotic relationship in the colon. Trillions of microorganisms colonize and survive in the colon without any overt overactivation of the host immune system, indicating that this mutual relationship involves well–controlled tolerogenic phenotype of the mucosal immune system. IDO1 is expressed at high levels in colonic epithelial cells as well as in mucosal immune cells under normal physiologic conditions, suggesting potential participation of the IDO1-mediated tryptophan catabolism pathway in the maintenance of immunotolerance against the commensal microorganisms in the colon. Recent studies have identified specific strains of commensals in the colon that generate tryptophan metabolites as potent agonists of AhR (18, 19). A variety of Lactobacilli in the colon produce indole-3-aldehyde from tryptophan catabolism that functions as an effective agonist for AhR in mucosal immune cells, resulting in the secretion of IL-22. This cytokine targets colonic epithelial cells to maintain the epithelial barrier and also to secrete antimicrobial peptides, both of which prevent the colonization and invasion of pathogenic bacteria in the colon. Similarly, certain fungal organisms are also capable of inducing IDO1 in mucosal epithelial cells such as in the female reproductive tract, resulting in the production of kynurenine from tryptophan catabolism, which then serves as the agonist for AhR in mucosal immune cells to create immunotolerance enabling the coexistence of these organisms with the host.

Tryptophan-Selective Amino Acid Transporter and Its Relevance to IDO1

The entire regulatory pathway involving IDO1 is obligatorily dependent on effective depletion of intracellular and extracellular tryptophan in the vicinity of IDO1-expressing immune cells as well as on the production of sufficient levels of kynurenine to activate AhR. This process has to involve a mechanism for the delivery of extracellular tryptophan to the intracellular enzyme IDO1. Tryptophan cannot traverse biological membranes by diffusion due to its hydrophilic nature, thus necessitating need for transporters in the plasma membrane. Mammalian cells express several transporters in the plasma membrane capable of transporting tryptophan, but almost all of them also transport other amino acids (20, 21). Since the extracellular medium bathing the immune cells in vivo contains all amino acids, selective depletion of tryptophan in the extracellular milieu is not likely unless the transporter is not only selective but also exhibits extraordinarily high affinity for tryptophan. Such a tryptophan-selective transport system exists in macrophages, dendritic cells and colonic epithelial cells, all of which are known to express IDO1 (22-24). This transport system accepts only tryptophan as the substrate with high affinity. More importantly, the same signals that induce IDO1 also upregulate the expression of the transport system. This coordinated regulation of the delivery system and the catabolizing enzyme in the same cell is critical for effective depletion of tryptophan both inside and outside the cell to initiate downstream signaling pathways.

Conclusion

The tryptophan-catabolism pathway involving IDO1 has profound impact on a broad spectrum of physiological and pathological conditions in health and disease. As such, the various components of this pathway potentially offer actionable therapeutic targets for various diseases. Since the pathway represents an immunosuppressive mechanism, pharmacological inhibition of this pathway has therapeutic potential in clinical and pathological conditions that require activation of immune function as a treatment; such a strategy might be effective, for example, in enhancing the ability of the host immune system to fight against cancer. IDO1 is a prime example; inhibitors of IDO1 are in clinical trials for cancer treatment. There may also be situations where activation of the pathway might be of clinical benefit; for example, autoimmune disorders and organ transplantation where suppression of immune function through activation of the tryptophan-catabolism pathway might be therapeutically effective. Each of the molecular components of the pathway is potentially exploitable as a drug target. An often-overlooked component of the pathway is the transport process in the plasma membrane of IDO1-expressing cells that delivers tryptophan to IDO1 for degradation. This transporter is obligatory for the entire pathway to function, thus representing an ideal target for development of drugs to inhibit or activate the pathway. Nonetheless, current efforts in targeting the tryptophan-catabolism pathway for therapeutic purposes are almost exclusively focused on IDO1. One of the primary reasons for the apparent lack of interest in the tryptophan-selective transporter as a therapeutic target is that neither the protein nor the gene responsible for the transport process has been identified at the molecular level. But pharmacological targeting of the transporter does not necessarily have to depend on the molecular identity of the transporter. The transport process is known at the functional level and its functional features have been established. For example, IFN-γ-inducible transport function that is selective for tryptophan can be used as a read out for the transporter in the design and development of pharmacological agents to modulate the activity of the pathway. Of course, information on the structure of the transporter protein would facilitate the discovery process in terms of modeling potential inhibitors or activators of the transporter. Given the importance of the transporter to the pathway as whole, continued efforts are needed to establish the molecular identity of the transporter.

References

[1] Geng, J. & Liu, A. (2014) Heme-dependent dioxygenases in tryptophan oxidation. *Arch. Biochem. Biophys.* 544: 18-26.

[2] Munn, D. H., Zhou, M., Attwood, J. T., Bondarev, I., Conway, S. J., Marshall, B., et al. (1998) Prevention of allogeneic fetal rejection by tryptophan catabolism. *Science* 281: 1191-1193.

[3] Popov, A. & Schultze, J. L. (2008) IDO-expressing regulatory dendritic cells in cancer and chronic infection. *J. Mol. Med. (Berl.) 86*: 145-160.

[4] Munn, D. H. & Mellor, A. L. (2013) Indoleamine 2,3-dioxygenase and metabolic control of immune responses. *Trends Immunol.* 34: 137-143.

[5] Prendergast, G. C., Metz, R., Muller, A. J., Merlo, L. M. & Mandik-Nayak, L. (2014) IDO2 in immunomodulation and autoimmune disease. *Front. Immunol.* 5: 585.

[6] Prendergast, G. C., Smith, C., Thomas, S., Mandik-Nayak, L., Laury-Kleintop, L., Metz, R. & Muller, A. J. (2014) *Cancer Immunol. Immunother.* 63: 721-735.

[7] Castilho, B. A., Shanmugam, R., Silva, R. C., Ramesh, R., Himme, B. M. & Sattlegger, E. (2014) Keeping the eIF2 alpha kinase Gcn2 in check. *Biochim. Biophys. Acta* 1843: 1948-1968.

[8] Fallarino, F., Grohmann, U. & Puccetti, P. (2012) Indoleamine 2,3-dioxygenase: From catalyst to signaling function. *Eur. J. Immunol.* 42: 1932-1937.

[9] Pallotta, M. T., Fallarino, F., Matino, D., Macchiarulo, A. & Orabona, C. (2014) AhR-mediated, non-genomic modulation of IDO1 function. *Front. Immunol.* 5: 497.

[10] Puccetti, P. & Grohmann, U. (2007) IDO and regulatory T cells: A role for reverse signaling and non-canonical NF-κB activation. *Nat. Rev. Immunol.* 7: 817-823.

[11] Orabona, C., Pallotta, M. T., Volpi, C., Fallarino, F., Vacca, C., Bianchi, R. et al. (2008) SOCS3 drives proteasomal degradation of indoleamine 2,3-dioxygenase (IDO) and antagonizes IDO-dependent tolerogenesis. *Proc. Natl. Acad. Sci. USA* 105: 20828-20833.

[12] Mezrich, J. D., Fechner, j. H., Zhang, X., Johnson, B. P., Burlingham, W. J. & Bradfield, C. A. (2010) An interaction between kynurenine and the aryl hydrocarbon receptor can generate regulatory T cells. *J. Immunol.* 185: 3190-3198.

[13] Opitz, C. A., Litzenburger, U. M., Sahm, F., Ott, M., Tritschler, I., Trump, S. et al. (2011) An endogenous tumour-promoting ligand of the human aryl hydrocarbon receptor. *Nature* 478: 197-203.

[14] Huang, K. & Fingar, D. C. (2014) Growing knowledge of the mTOR signaling network. Semin. *Cell Dev. Biol.* 36: 79-90.

[15] Abraham, R. T. (2015) Cell biology. Making sense of amino acid sensing. *Science* 347: 128-129.

[16] Metz, R., Rust, S., Duhadaway, J. B., Mautino, M. R., Munn, D. H., Vahanian, N. N., Link, C. J. & Prendergast, G. C. (2012) IDO inhibits a tryptophan sufficiency signal that stimulates mTOR: A novel IDO effector pathway targeted by D-1-methyl trylptophan. *Oncoimmunology* 1: 1460-1468.

[17] Cobbold, S. P., Adams, E., Farquhar, C. A., Nolan, K. F., Howie, D., Lui, K. O. et al. (2009) Infectious tolerance via the consumption of essential amino acids and mTOR signaling. *Proc. Natl. Acad. Sci. USA* 106: 12055-12060.

[18] Romani, L., Zelante, T., De Luca, A., Iannitti, R. G., Moretti, S., Bartoli, A., Aversa, F. & Puccetti, P. (2014) Microbiota control of a tryptophan-AhR pathway in disease tolerance to fungi. *Eur. J. Immunol.* 44: 3192-3200.

[19] Yanagihara, S., Goto, H., Hirota, T., Fukuda, S., Ohno, H. & Yamamoto, N. (2014) Lactobacillus acidophilus L-92 cells activate expression of immunomodulatory genes in THP-1 cells. *Biosci. Microbiota Food Health* 33: 157-164.

[20] Broer, S. (2008) Amino acid transport across mammalian intestinal and renal epithelia. *Physiol. Rev.* 88: 249-286.

[21] Ganapathy, V. (2012) Protein digestion and absorption. In: *Physiology of the Gastrointestinal Tract* (Ed., Johnson, L. R.), 5th edition, pp. 1595-1623. Elsevier.

[22] Seymour, R. L., Ganapathy, V., Mellor, A. L. & Munn, D. H. (2006) A high-affinity, tryptophan-selective amino acid transport system in human macrophages. *J. Leukoc. Biol.* 80: 320-327.

[23] Silk, J. D., Lakhal, S., Laynes, R., Vallius, L., Karydis, I., Marcea, C., Boyd, C. A. & Cerundolo, V. (2011) *J. Immunol.* 187: 1617-1625.

[24] Bhutia, Y. D., Babu, E. & Ganapathy, V. (2015) Interferon-γ induces a tryptophan-selective amino acid transporter in human colonic epithelial cells and mouse dendritic cells. *Biochim. Biophys. Acta* 1848: 453-462.

INDEX

D

E

H

I

J

K

L

M

N

O

P

Q

R

S

T

U

V

W

Y